Groundwater Assessment and Management
for Sustainable Water-Supply and Coordinated Subsurface Drainage

A Guidebook for Water Utilities and Municipal Authorities

Groundwater Assessment and Management for Sustainable Water-Supply and Coordinated Subsurface Drainage

A Guidebook for Water Utilities and Municipal Authorities

Stephen Foster and Radu Gogu

IWA Groundwater Management Specialist Group

Published by   **IWA Publishing**
Unit 104–105, Export Building
1 Clove Crescent
London E14 2BA, UK
Telephone: +44 (0)20 7654 5500
Fax: +44 (0)20 7654 5555
Email: publications@iwap.co.uk
Web: www.iwapublishing.com

First published 2022
© 2022 IWA Publishing

Disclaimer

British Library Cataloguing in Publication Data
A CIP catalogue record for this book is available from the British Library

ISBN: 9781789063103 (Paperback)
ISBN: 9781789063110 (eBook)
ISBN: 9781789063127 (ePub)

Contents

Preface

Groundwater beneath cities is important. Water utilities and private abstractors use is it as a secure source of water-supply, and municipal authorities have to cope with it when using underground space for buildings, facilities and transportation infrastructure. However, these agencies often pay little attention to groundwater once their wells are installed or subsurface construction is completed. This Guidebook aims to highlight what water utilities and municipal governments can do to improve groundwater assessment, management and monitoring to avoid many 'nasty surprises'.

Groundwater, especially from deeper aquifers, is a critical resource for enhancing urban water-supply security under climate change stress. But to achieve its use sustainably will require effective promotion of an appropriate level of adaptive resource management and protection, according to local circumstances. Moreover, groundwater conditions at shallower depths are an essential issue when it comes to sound sanitation planning.

In recent times, municipal governments have made much more use of urban subsurface space (especially down to 15-metres depth) for construction – notably of deep basements, vehicle parking spaces and transportation routes. Traditionally the drainage and stability of such structures was achieved through individual site investigation and design, but today a more coordinated approach is needed to manage shallow groundwater conditions. And these conditions are often aggravated by a rising shallow water table due to the urbanisation process itself and as a result of the impact of climate change.

This **Guidebook** is divided into three complementary parts:

- **Part A is intended primarily for the guidance of managers, engineers and scientists in water utilities, water resource agencies and municipal sanitation**, working to improve urban water-supply resilience, with its inevitable requirement to get more involved in groundwater management. Part A is divided into two sections, which deal respectively with undertaking essential diagnostic procedures and with formulating strategic actions.
- **Part B is intended primarily for guidance of engineers, planners and managers in municipal government authorities** working to improve the design and execution of subsurface urban infrastructures to avoid potentially-costly subsurface drainage issues, structural instability and groundwater flooding problems arising from shallow and/or rising water tables. Part B is divided into two sections which deal respectively with the characterisation of problems and with essential steps in taking more integrated action.
- **Part C presents a series of 6 city case histories of groundwater management actions,** identifying the key issues that needed attention in each case and the institutional arrangements that facilitated action being taken.

Acknowledgements

Developing the concept and executing the drafting of this Guidebook on Groundwater Management was primarily the work of the principal authors, Stephen Foster (UK) and Radu Gogu (Romania), who have served during 2018-22 as the Chair and Deputy Chair of the IWA-Groundwater Management Special Group respectively.

All members of the Groundwater Management SG were involved to some extent in its production, and the authors wish to give a generous acknowledgement to Michael Eichholz (Germany), Ricardo Hirata (Brasil), Faiz Alam (India) and Julia Gathu (Kenya and SG Secretary) for their major contribution, and also to Susie Mielby (Denmark), Aleksandra Tubic (Serbia) and Ayoade Agbite (Nigeria) for their significant inputs.

The authors also wish to thank Rachna Sarkari of the IWA-Headquarters Team (and her predecessor Kambiri Cox) for encouragement on the concept of the Guidebook, and on very efficient liaison on the procedures for IWA publication in 2022, the UN Year of Groundwater. Mark Hammond is also thanked for his work on publishing the manuscript.

doi: 10.2166/9781789063110_0001

Part A

Groundwater for Urban Water-Supply

A1 ESSENTIAL DIAGNOSTIC PROCEDURES

Why is groundwater important to water utilities?

Wherever available, groundwater resources have significant advantages as a primary source of water utility supply since:
- they provide a climate-resilient source of water-supply because of the large natural storage of groundwater systems; and
- they facilitate flexible stepwise water-supply development, in response to growth in population and per capita water demand.

However, if groundwater resources are not managed adequately they can be degraded by a number of processes including:
- uncontrolled access by private waterwell users, which can degrade the aquifer system and diminish water utility revenue considerably; and
- gradual deterioration of groundwater quality due to inappropriate land use, industrial discharges to the ground, wastewater seepage from on-site sanitation and leaking sewers, and from other sources.

How can groundwater impact water service utilities?

There are a number of diverse ways in which groundwater can greatly benefit or seriously prejudice water utility operations. It is thus essential to have an adequate understanding of the relevant mechanisms and interactions from the outset.

Groundwater systems interact with various facets of the urban water cycle in a number of ways, shown schematically in Figure A1. This diagram reveals the hidden, but intimate, relationship between water utility operations, private waterwells and the underlying groundwater system. Among the most important groundwater system interactions are the following:
- they provide a significant component of the total water-supply
- they allow uncontrolled access for private waterwell users, thus diminishing potential water utility revenue significantly
- they generate a significant inflow to main sewerage systems from 'shallow water tables', thus causing additional treatment burdens.

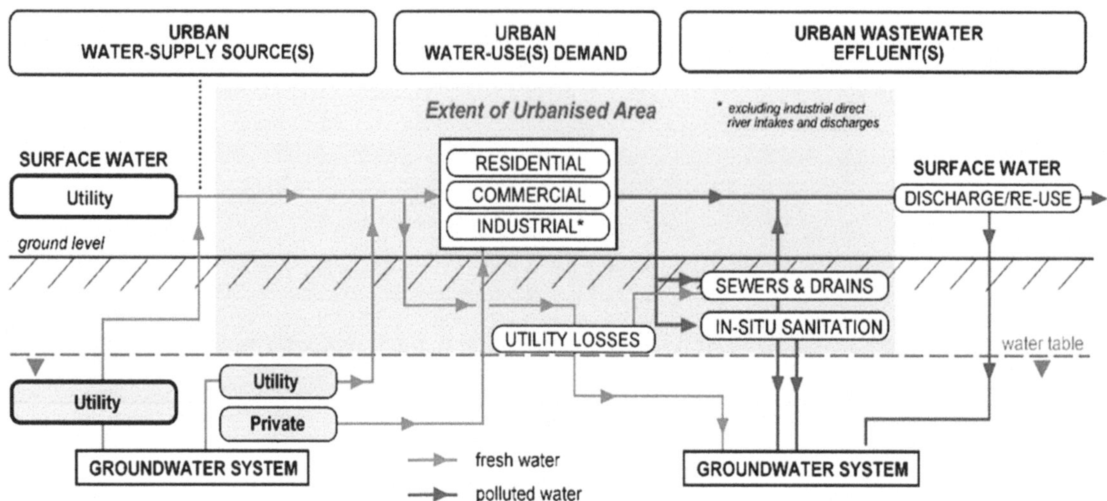

Figure A1 Typical scheme of interaction between water utility operations and a groundwater system (after Foster 2020a).

Extraction of groundwater by both water utilities and private users tends to evolve over decades (Figure A2), with deficiencies in public water-supply availability (which can in some instances be large) being met by increased private waterwell use.

The effects of climate change on groundwater are complex and spatially variable. In the longer run, recharge rates and quality are likely to be affected, in addition to changes in groundwater demand and use. While it will be important to have an appraisal of the relevant processes and their potential impact, this should not overshadow the fact that groundwater resources will in many cases be the intermediate to long-term solution for climate change adaptation, given the very large storage resources of most aquifers.

It should be noted, however, that the processes of urbanisation, and the major change in land use that they imply, themselves generate important changes in the groundwater system which will need to be confronted (IUCN, 2016):

- varying aquifer recharge rates, with a general tendency to significant increase, since in many cases recharge from leaking water mains and/or *in situ* sanitation discharge more than counterbalances reduction due to land surface impermeability.
- declining recharge quality due to a variety of urban groundwater pollution hazards (Figure A3), most notably *in situ* sanitation and industrial/commercial discharge of effluents to the ground.

What data are needed to diagnose groundwater management needs?

If a water utility obtains a significant part of its total water-supply from groundwater it is essential to have detailed knowledge of:

- the number of utility waterwells and/or springs in use, their current maximum capacity, and their design and distribution (intra-urban, external wellfields, etc.)
- the current mode of deployment of groundwater sources (for baseload supply, use conjunctively with surface water sources, supply of specific or difficult zones only, etc.).

It will also be necessary to have a sound understanding of:

- the class of groundwater system occurring within the water utility operational area (alluvial, sedimentary, karstic or bedrock aquifers)

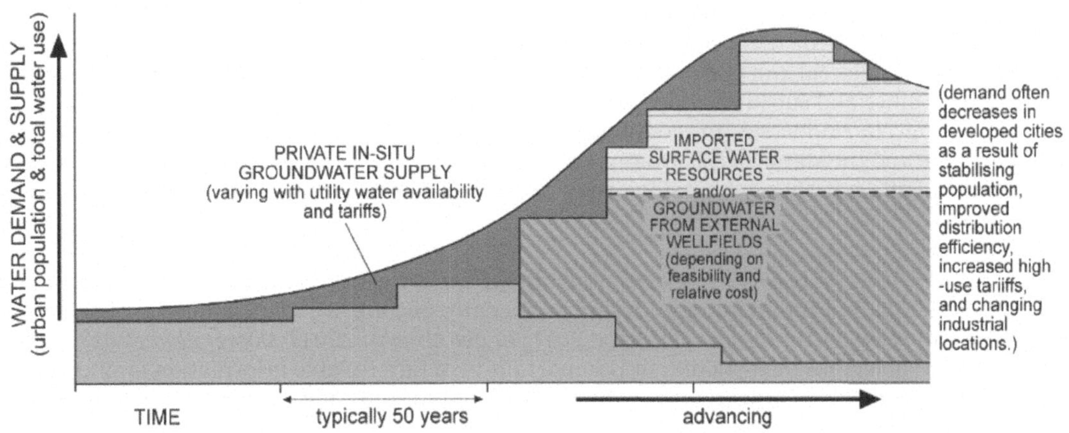

Figure A2 Typical evolution of groundwater use and dependency with urban population growth.

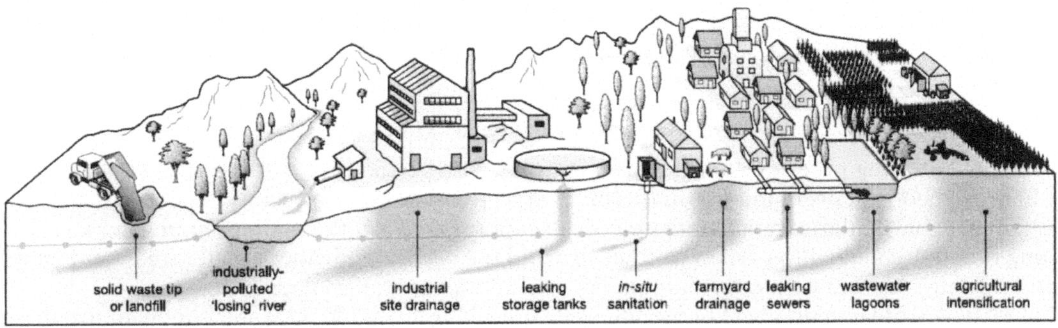

Figure A3 Land use activities commonly generating a groundwater pollution hazard.

- the current status of groundwater resources and any evidence of a long-term falling water table due to local aquifer over-exploitation
- the present groundwater quality situation and whether there is evidence of significant pollution, from industrial or agricultural sources, or from *in situ* sanitation or other urban infrastructure, or natural geogenic contamination.

The following also require careful technical appraisal:
- the extent to which mains sewer collectors are installed below the water table and the increase in total sewer flow that arises as a result
- the extent to which contributory sewers are perched above the water table, with any seepage potentially causing groundwater pollution
- whether local sanitation is by *in situ* systems, and whether these involve high water use or are essentially 'dry systems' with urine separation.

Groundwater quality concerns and pollution hazards will be a particularly important consideration where water utility wells are located within urban and industrialised areas. As a result many water utilities are progressively moving their groundwater abstraction to wellfields outside urbanised areas, and taking action to secure significant control of land-use activity in the wellfield groundwater capture area.

Monitoring of groundwater quality is an essential activity for water utilities, and the scope of this may have to be broadened to cover (at least occasionally) some emerging pollutants (derived from community pharmaceutical and personal-care products, food additives, etc.), which are demonstrating considerable mobility and subsurface persistence.

In summary it is important to commission systematic stocktaking of current water utility groundwater use and the status of the groundwater resource and quality in the aquifer(s) on which the utility depends. Different physical typologies will require different management responses.

How is private urban waterwell use relevant to water utilities?

Construction of private waterwells on residential properties in the urban areas of many cities in South Asia, Latin America and Sub-Saharan Africa has recently become a 'boom industry' (Foster *et al.*, 2010d; Gronwall *et al.*, 2010; Foster & Hirata, 2011; Lapwoth *et al.*, 2017; Foster *et al.*, 2018; Alam & Foster, 2019). In such cities in particular, it is important to obtain reliable information on the numbers of private wells operating in and around a water utility supply zone, because of their potential impact on water utility operations (Table A1).

Of special significance are the types of water use involved, and whether their existence represents:
- a significant loss of potential revenue from water utility sales
- a cause of groundwater stress and competition for limited available resources
- a burden on mains sewerage systems where present, recognising that part of the wastewater generated has not been supplied by the water utility and thus no sewerage charge has been levied.

Could water utilities have 'hidden interests' in groundwater management?

Agreements outlining groundwater management plans for key aquifer systems should be associated with improved control over land-use in their recharge areas and a significant reduction in subsurface pollution load, particularly in the land area of the capture zones of the main groundwater sources. In the long run, this should allow water utilities to avoid the need for advanced water treatment facilities for groundwater sources with a resultant major capital and operational cost saving.

The existence of large numbers of private wells in use for independent water-supply in urban areas inevitably results in a major loss of revenue from water sales for the corresponding water utility.

Table A1 Overview of the impact of large-scale urban use of private waterwells.

<table>
<tr><td colspan="2" align="center">Urban Private Residential Waterwells</td></tr>
<tr><td>Pros</td><td>Cons</td></tr>
<tr>
<td>
<ul>
<li>improves access and reduces costs for some user groups (but not for the poorest unless capital investment is underwritten or the water table is shallow)</li>
<li>especially appropriate for 'non quality-sensitive' uses, which could be stimulated to reduce pressure on municipal supply</li>
<li>reduces pressure on municipal supply and can be used for demands whose location or peak levels present difficulty</li>
<li>can recover a significant proportion of water mains leakage</li>
</ul>
</td>
<td>
<ul>
<li>interactions with in situ sanitation could cause a public health hazard and make any waterborne epidemic difficult to control</li>
<li>hazardous where natural groundwater contamination is present</li>
<li>may encounter sustainability problems in cities where the principal aquifer is significantly confined and/or mains water-supply leakage is low</li>
<li>inefficient development where a large number of private water wells are constructed within a small area rather than a single highly productive installation</li>
<li>distorts the technical and economic basis for water utility operations with implications for utility investment and tariffs</li>
</ul>
</td>
</tr>
</table>

Any rationalisation of groundwater use (especially the elimination of private wells for self-supply where an adequate main water-supply is available) that can be achieved through the implementation of an urban groundwater management plan should result in additional revenue from water sales for the corresponding water utility.

Who has the institutional responsibility for groundwater resources?

The commonest situation is one where a water resource regulatory agency exists (at national and/or regional level) and nominally has lead responsibility for groundwater resource management (Foster & Vairavamoorty, 2013). However, it is also important to assess whether they have sufficient on-the-ground capacity to be effective in regulating groundwater use and in protecting groundwater quality against pollution in the water utility operational zone.

Moreover, it is also important to know:
- whether there is an established mechanism for regular dialogue on groundwater issues between the regulatory agency and water utility
- whether the water utility has a clearly defined role for some component of groundwater management, such as the provision of regular monitoring data on groundwater levels and quality.

There may be additional actual or perceived impediments to a water utility becoming involved with groundwater management (Table A2). The water utility should consider undertaking a systematic analysis of the potential groundwater stakeholders in their operational area, and assessing the scope for forming productive partnerships with other agencies and for sharing the cost of groundwater management activities (FAO-UN, 2016).

Does your water utility have capacity for groundwater management?

Relevant criteria here are the number of staff employed by your water utility on groundwater-related issues, and their professional background and technical training. They may only be competent in relation to well operations, or may also have some training in hydrogeological investigation and groundwater resource evaluation.

Given suitable staff, a water utility may already have undertaken an independent assessment of aquifer pollution vulnerability, surveyed subsurface contaminant loads and evaluated groundwater pollution hazard.

Table A2 Factors impeding water utility involvement with groundwater management.

Factor	Outcome
Utility assumption (sometimes reinforced by legislation) that groundwater resource management and protection are the sole responsibility of another organisation	Responsibility entrusted entirely to an environment agency, water resources ministry or basin authority
Utility perception that 'safe drinking-water quality' cannot be achieved by groundwater protection measures	Presumption that the required quality can only be guaranteed by advanced water treatment (with cost charged to water users)
Utility operating under a time-limited concession to a public body (municipal authority or national ministry) which requires only a reduction of mains leakage/ unaccounted-for water	Development of new groundwater sources and their protection through land-use management agreements is outside of the utility's remit
Utility has to conform with pre-defined local geo-political boundaries in its operations as prescribed under municipal concession	Seriously constrains approach taken to wellfield construction, aquifer management and protection
Utility size is too small to allow it to contribute to groundwater management and protection	Insufficient authority over required land area to be effective

Alternatively, the water utility may have brokered a long-term relationship with a local university department, water resources institute, geological survey or specialist consultancy for the provision of dedicated services in the area of groundwater investigation, assessment and policy.

Whatever the position of your own utility, it is important that senior managers are in possession of a complete diagnostic assessment of the utility's dependence on groundwater resources, current groundwater quality status and trends, and of the other stakeholders in urban groundwater and their impact on its quality. This knowledge will be critical in any liaison with the local water resource agency and municipal sanitation department over the sustainable management of groundwater resources for public water-supply.

A2 FORMULATING STRATEGIC ACTIONS

Most water utilities with a degree of dependence on groundwater would benefit greatly from drawing up a 'strategic action plan' for the management and protection of the groundwater resources on which they depend (IAH, 2015; IAH, 2019). This action plan should have a number of components (further elaborated in Table A3) which broadly address:
- groundwater resource management
- groundwater quality protection
- groundwater monitoring for adaptive management.

What actions are needed as regards groundwater resource management?

Understanding the resource

A sound knowledge of the groundwater system(s) used by the water utility is an essential basis for effective resource management. Since the hydrogeological characterisation of a groundwater system is a complex task needing specialist knowledge and financial resources, water utilities rarely engage in this task alone and should seek cooperation with water resource agencies, research institutes, geological surveys and universities.

It will be necessary to assess groundwater flow directions, recharge areas and rates, and use the results to develop a conceptual groundwater model of the system(s) involved. If sufficient monitoring data are available, a numerical groundwater model can be developed to permit analysis of crucial parameters and trends, and to build potential management scenarios.

However, sophisticated technical tools are of little use without sound understanding of the groundwater system, and for water utilities it is important to participate actively in the assessment of groundwater systems, as a means of fully understanding the processes involved.

Controlling abstraction

Sound groundwater management requires the rational use of the resource, and a water utility has an important task here in fostering rational and efficient water use. The utility should make efforts to **optimise its own groundwater production wells**, including regular pump **maintenance** both to save energy and prolong waterwell and pump life.

Efficient use of the groundwater produced is another crucial step, involving the **reduction of distribution system water losses** and encouraging consumers to install water-saving appliances. A **tariff system** that provides **incentives for water saving** with universal access, social tariffing and cost recovery is required.

Water utilities also have responsibility for **integrating the use of a diverse range of water sources,** which helps to reduce the impact of drought and improve water-supply security under climate change pressures.

Water utilities should have a strong interest in the groundwater system around their service area, and also move to ensure that **waterwell construction by private households or businesses is limited**.

Table A3 Overview of components of a 'strategic action plan'.

Understanding the Resource	System Assessment • groundwater flow directions • aquifer recharge areas and rates Groundwater Model • conceptual modelling and scenario building • numerical modelling
Controlling Abstraction	Utility Waterwell Management • waterwell maintenance • efficiency testing/improvement Water Distribution Efficiency and Loss Reduction Rationalising the Approach to Private Water Wells
Strategic Monitoring	Groundwater Monitoring System Design • abstraction metering • water-level measurement • quality investigation and surveillance • data management and exchange
Quality Protection	Point-Source Pollution Inventory • source categorization and control • emergency planning Diffuse-Source Pollution Assessment and Control • mechanism for agricultural sector liaison • land-use regulation possibilities Municipal Wastewater and Solid Waste Planning • promote prioritisation of sewered sanitation • influence decisions on landfill locations Groundwater Protection Zone Definition • pre-treatment of specific industrial effluents • farmer cooperation to reduce diffuse pollution • purchase or control over critical land areas
Organisation of Actions	Definition of Responsibilities • creation of groundwater management unit with training • formally establish key contacts and exchanges Promotion of Strategic Action Plan • lobbying for groundwater management and protection • undertake public awareness campaigns • generate networks for water stewardship

A management strategy to deal with this practice should be based on two principles – promoting the regulation of private well drilling and addressing the underlying causes of private water well use.

There is a strong argument that a water service utility should be a statutory consultee of the water resource agency as regards the issuing of licences to construct and operate private waterwells in urban areas, and that the water utility should use this process to object (a) should a proposed well interfere with their groundwater sources and (b) if they have ample water-supply availability in the area concerned. Moreover, if a private well is designated for commercial production and/or distribution of a water-supply, special payments should normally be levied. Even if private domestic waterwells are exempt from need for a licence they should be registered, and the water utility should simultaneously have the legal right to levy a sewerage charge in respect of the abstraction of a private waterwell if this generates a sewer flow (often assumed to be 75% of the private well production capacity).

Organisation of actions

Water utilities need to define clearly roles and responsibilities as regards groundwater management and protection. Internally a team of dedicated staff should be responsible for groundwater management issues, reporting directly to a high level in the organisation.

These staff require some training in hydrogeology to translate scientific findings into recommendations that are understandable by other departments and decisions makers. They also require institutional skills to lobby and convince other stakeholders to protect the resource, and to find innovative cooperation agreements and legal provisions. Public relations and communications skills will be required to promote the implementation of groundwater protection measures with land owners, farmers, businesses and industries.

Water utilities need to build alliances and form networks that bring the key stakeholders of groundwater management and protection together, through water stewardship approaches or multi-stakeholder platforms.

What actions are needed as regards groundwater quality protection?

Context for action

Groundwater systems worldwide are experiencing an increasing threat of pollution from agricultural land-use, urbanisation, industrial development and mining enterprises. All too frequently, those who depend on such resources for the provision of potable water supplies have taken no significant action to assure raw water quality, nor have they made adequate efforts to assess potential pollution hazards.

Proactive campaigns and practical action to protect the (generally excellent) natural quality of groundwater are widely required. It is important that water utilities make assessments of the strategic value of their groundwater sources, based on a realistic evaluation of replacement value, which includes both the cost of developing a new source and connecting that potentially distant source into existing distribution networks.

Groundwater pollution hazard assessments are needed to provide a clearer appreciation of actions needed to protect groundwater quality and, if undertaken by the water utility, they should prioritise both preventive action to avoid future pollution and corrective actions to control existing threats. Cooperation with municipal sanitation departments is important to influence decision making on priority areas for sewered sanitation and the siting of solid waste landfills.

For potable mains water-supply, a high and stable raw water quality is a prerequisite, and one that is best met by protected groundwater sources. Recourse to treatment processes to achieve this (beyond precautionary disinfection where fecal contamination is a significant threat) should be regarded as a last resort, in view of technical complexity and financial cost, and the operational burden they impose.

Understanding pollution processes

Most groundwater originates as excess rainfall infiltrating (directly or indirectly) at the land surface. In consequence, activities at the land surface can threaten groundwater quality. The more common types of activity capable of causing significant groundwater pollution and the most frequently encountered contaminants are summarised in Table A4; they depart widely from the activities and compounds commonly polluting surface water bodies. Certain activities (and specific processes or incremental practices) often present disproportionately large threats to groundwater quality, and thus sharply focused and well-tuned pollution control measures can produce major benefits for relatively modest cost.

The potential contaminant load can be classified in a simplified but pragmatic way – whether it is essentially point or diffuse source, according to the types of products involved, and their mode of entry into the subsurface – and enables the greatest hazards to be identified.

Natural subsoil profiles attenuate many water pollutants and have long been considered potentially effective for the safe disposal of domestic wastewater. The elimination of contaminants in the vadose (unsaturated) zone is the result of biochemical degradation and chemical reaction,

Table A4 Common sources and types of groundwater pollution in urban areas.

Pollution Source	Typical Contaminants
In-situ Sanitation	Nitrates; halogenated hydrocarbons; microorganisms
Gas Stations and Garages	Aromatic hydrocarbon; benzene; phenols; halogenated hydrocarbons
Solid Waste Disposal	Ammonium; salinity; halogenated hydrocarbons; heavy metals
Metal Industries	Trichloroethylene; tetrachloroethylene; halogenated hydrocarbons; phenols; heavy metals; cyanide
Painting and Enamel Works	Alkylbenzene; halogenated hydrocarbons; metals; aromatic hydrocarbons; tetrachloroethylene
Timber Industry	Pentachlorophenol; aromatic hydrocarbons; halogenated hydrocarbons
Dry Cleaning	Trichloroethylene; tetrachloroethylene
Pesticide Manufacture	Halogenated hydrocarbons; phenols; arsenic
Sewage Sludge Disposal	Nitrates; halogenated hydrocarbons; lead; zinc
Leather Tanneries	Chromium; halogenated hydrocarbons; phenols

but contaminant retardation due to sorption is also of importance increasing the time available for contaminant elimination. However, not all subsoil profiles and underlying strata are equally effective in contaminant attenuation and groundwater systems will be particularly vulnerable to pollution where, for example, consolidated highly fissured rocks or highly permeable sediments are present. The degree of attenuation will also vary widely with types of pollutant and polluting process in any given situation.

Concern about groundwater pollution relates primarily to the so-called unconfined groundwater systems, especially where their vadose zone is thin and the water table shallow, but a significant pollution hazard may also be present, even where groundwater is semi-confined, if the confining aquitards are relatively thin and permeable.

Water movement and contaminant transport from the land surface to groundwater systems can be a very slow process, and it may take decades before the impact of a pollution episode by a persistent contaminant becomes fully apparent in groundwater supplies from deeper wells. This can simultaneously be a valuable benefit and a serious concern because it allows time for the breakdown of degradable contaminants but may lead to complacency about the likelihood of penetration of persistent contaminants. The implication is also that once groundwater quality has become polluted, large volumes are usually involved – thus clean up measures nearly always have a high economic cost and are often technically problematic.

Assessing groundwater pollution hazards

The term 'groundwater pollution hazard' relates to the probability that groundwater in an aquifer will become contaminated to concentrations above the corresponding WHO guideline value for drinking-water quality. The most logical approach to a groundwater pollution hazard is to regard it as the interaction between:

- an aquifer's pollution vulnerability consequent upon the natural characteristics of the strata separating it from the land surface; and
- the contaminant load that is – or might be – applied on the subsurface environment as a result of human activity.

The term 'aquifer pollution vulnerability' is intended to represent sensitivity of a groundwater system to being adversely affected by an imposed contaminant load (Table A5), and can be readily mapped (Foster *et al.*, 2002; Foster *et al.*, 2013). On such maps, the results of surveys of potential

Table A5 Practical interpretation of classes of aquifer pollution vulnerability.

Vulnerability Class	Corresponding Definition
Extreme	vulnerable to most water pollutants with rapid impact in many pollution scenarios
High	vulnerable to many pollutants (except those strongly absorbed or readily transformed) in many pollution scenarios
Moderate	vulnerable to some pollutants but only when continuously discharged or leached
Low	only vulnerable to conservative pollutants in the long term when continuously and widely discharged or leached
Negligible	confining beds present with no significant vertical groundwater flow (leakage)

subsurface contaminant load can be superimposed to facilitate the assessment of any groundwater pollution hazard, and an assessment of hazard to a specific groundwater supply can be undertaken by superimposing the groundwater pollution hazard maps produced onto a map of the groundwater capture perimeters of the source concerned.

Designing groundwater protection plans

The role of water utilities in groundwater quality protection programmes, and their relationship with the other main stakeholders, is illustrated in Figure A4. A sensible balance needs to be struck between the protection of groundwater resources (aquifers as a whole) and specific sources (boreholes, wells, and springs). While both approaches to groundwater pollution control are complementary, the emphasis placed on one or the other will depend on the resource development situation and on the prevailing hydrogeological conditions.

If potable use comprises only a minor part of the total available groundwater resource, then it may not be cost-effective to protect all parts of an aquifer equally. Source-oriented strategies will then be

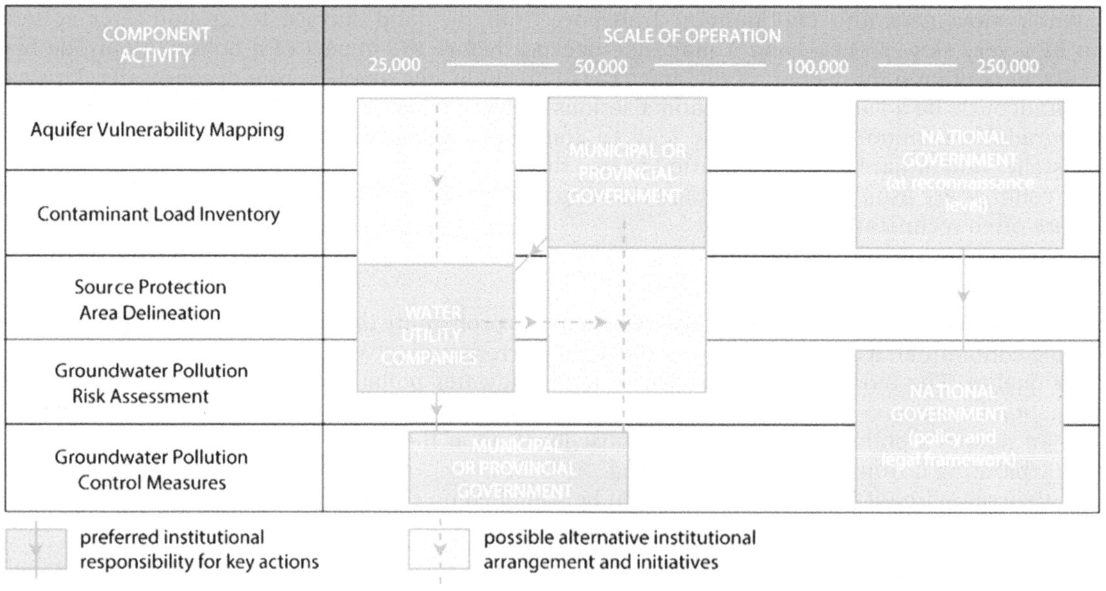

Figure A4 Role of water utilities in the general scheme of groundwater quality protection.

Table A6 Criteria for the delineation of groundwater source protection zones (SPZs).

Protection Zone	Criteria for Delineation
Source Recharge Capture Zone	The zone in which all recharge (whether from precipitation or surface watercourses) will be captured, determined in area by water balance considerations and in geometry by groundwater flow paths (and can be located distant from the source in confined aquifers). In extended drought, the actual capture zone will be larger than the area protected
Microbiological Protection Zone	The distance equivalent to a specified horizontal flow time in a saturated aquifer. In all reported waterborne-disease outbreaks, the proven source of pollution was less than 20 days of groundwater flow, but some hardy pathogens are capable of subsurface survival for 400 days – thus a 50-day isochron is widely taken as a reasonable parameter
Wellhead Operational Zone	The small area around a source, preferably owned by the groundwater abstractor. The only permitted activities in this area are related to water abstraction. It should have a concrete floor to prevent infiltration of oils and chemicals during maintenance. A radius of at least 20 metres is highly desirable with inspection of sanitary integrity undertaken over a 200 metre radius

appropriate and will involve work at scales in the range 1 : 25 000 to 1 : 100 000, commencing with the delineation of the groundwater capture area of water-supply sources (Table A6 and Figure A5), and then including assessment of aquifer pollution vulnerability and subsurface contaminant load in the areas defined.

This approach is best suited to more uniform, unconsolidated aquifers exploited by a relatively small and fixed number of high-yielding municipal water-supply boreholes with stable pumping regimes. It is most appropriate in less densely populated regions where their delineation can be conservative without producing conflict with other interests. They cannot be so readily applied where there are very large and rapidly growing numbers of individual abstractions, which render consideration of individual sources and the establishment of fixed zones impracticable.

Aquifer-oriented strategies are more universally applicable, since they endeavour to achieve a degree of protection for the entire groundwater resource. They begin with aquifer pollution vulnerability mapping of more extensive areas (including one or more important aquifers) working at a scale of 1 : 50 000 to 1 : 100 000. Such mapping would normally be followed by an inventory of subsurface contaminant load at a more detailed scale, at least in the more vulnerable areas.

To protect aquifers against pollution it is necessary to constrain – both existing and future – land-use, effluent discharge, and waste disposal practices. Possible methods of control of common sources of potential groundwater pollution are given in Table A7.

It is possible to manage land entirely in the interest of groundwater abstraction, and there are an increasing number of European examples of water-supply companies owning virtually entire recharge areas to prevent microbiological contamination of groundwater supplies (Thomsen *et al.*, 2004). But this may not be acceptable on socioeconomic grounds, and it is then necessary to define protection strategies that accept trade-offs between competing interests. An increasingly-used strategy is to negotiate a cooperation arrangement between the water utility and farmers within their groundwater source protection zones (SPZs), with less intensive cultivation practices (restricting fertiliser and pesticide applications, and livestock grazing densities) in exchange for compensation payments.

Instead of applying universal controls over land-use and effluent discharge to the ground, it is less prejudicial to economic development to utilise the natural contaminant attenuation capacity of the strata overlying the aquifer when defining the level of control necessary to protect groundwater quality. Simple and robust zones (based on aquifer pollution vulnerability and source protection perimeters) need to be established (Figure A6), with matrices to indicate what activities are possible and where they are acceptable or unacceptable for groundwater (Table A8).

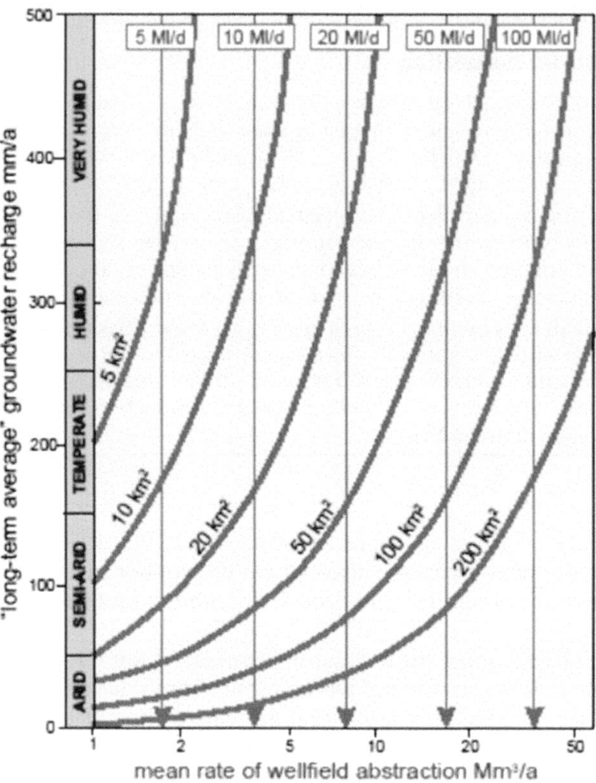

Figure A5 Variation in land area needed for complete protection of a groundwater or wellfield source.

Table A7 Methods of control of selected common sources of groundwater pollution.

Pollution Type	Pollution Source	Possible Restrictions	Alternative Methods
Diffuse Source	Agricultural fertilisers and pesticides	Strict management to meet crop needs, control of timing/rate of application, ban on selected pesticides	None
	In situ sanitation (latrines, cesspits, septic tanks)	If water use is high, choose septic tank, adhere strictly to design standards	Mains piped sewerage
Point Source	Underground storage tanks and pipelines	Double lining	Install above ground with leak detection
	• Solid waste disposal • domestic • industrial	Domestic: base impermeabilisation Industrial: leachate collection with treatment/recycling	Remote disposal
	Effluent lagoons (agricultural, municipal, industrial)	Base impermeabilisation	None, except treatment plant and remote disposal
	Cemeteries	Tomb impermeabilisation	Crematoria

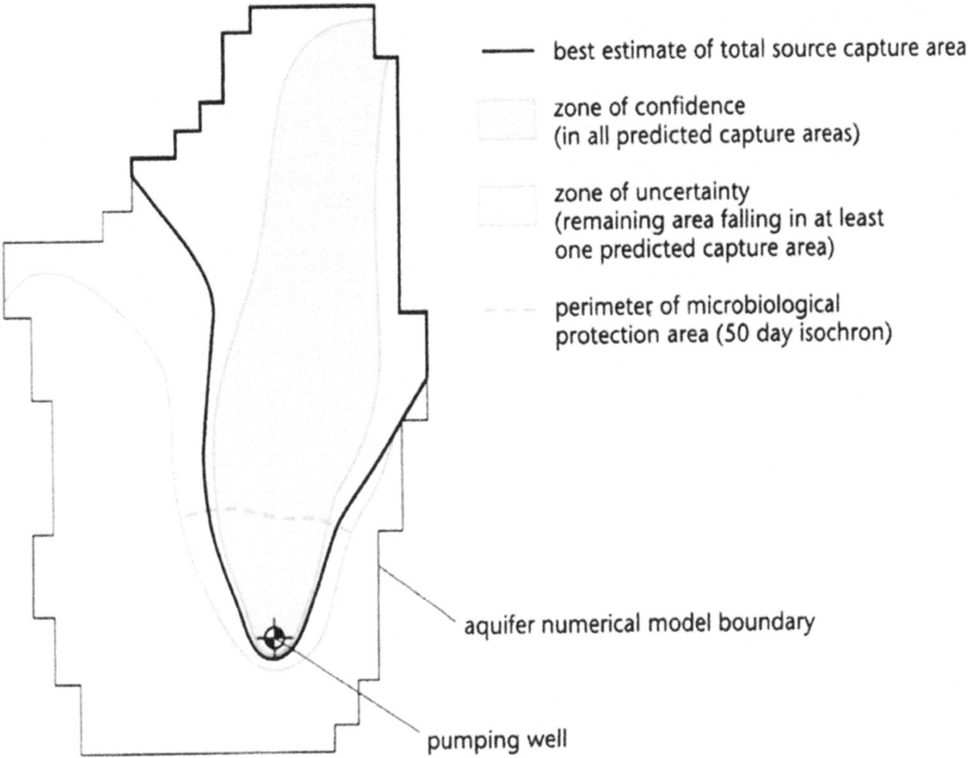

Figure A6 Numerical aquifer modelling for SPZ delineation incorporating zones of confidence and uncertainty.

There is an overriding case for land surface zoning as a general framework for the promotion of groundwater protection policy because:

- decisions will be made affecting groundwater in any event, and if planners have no zoning guidance this will lead to less consultation with those concerned with water resources
- it is unrealistic to expect exclusive protection for all groundwater, and a zoning strategy is important to ensure that trade-offs between economic development and aquifer protection are made objectively.

Groundwater protection zoning also has a key role in setting priorities for groundwater quality monitoring, the environmental audit of industrial premises, pollution control within the agricultural advisory system, and in public education generally. These activities are essential components of a comprehensive strategy for groundwater quality protection.

Urban sanitation and wastewater planning

The provision of sanitation services for fast-growing urban centres is a major global engineering challenge. Effective and universal urban sanitation is fundamental to healthy and productive human life and the composition of wastewater makes it a potentially serious groundwater contaminant. However, properly managed, it can also be regarded as a useful resource. A conceptual source–pathway–target approach can be used as the basis for appraising the groundwater pollution hazard arising from wastewater in urban settings.

Table A8 Acceptability of potentially polluting activities based on the aquifer vulnerability and special protection zone approaches.

POTENTIAL CONTAMINANT LOAD	AQUIFER POLLUTION VULNERABILITY ZONES *			GROUNDWATER SOURCE PROTECTION AREAS	
	low	medium	high	500-day	50-day
reduced	3	3	2	2	1
moderate	2	2	1	1	1
elevated	2	1	1	1	1

ACTION-LEVEL
1 = high　　　2 = intermediate　　　3 = low

* Numbers of zones/areas reduced to simplify presentation.

While effective sanitation is fundamental for human health, cities of all sizes face a growing challenge in providing appropriate systems as a result of limited institutional and financial capacity, and insufficient political will. In developing countries, much urban sanitation is often provided by *in situ* sanitation systems (septic tanks, dry or pour-flush latrines). Even assuming that the subsoil conditions are favourable for such sanitation units (and vadose zone filtration is effective in eliminating fecal pathogens), their utilisation at high population density usually results in serious pollution of shallow aquifers by nitrates and community chemicals. Thus groundwater pollution hazard has to be an important consideration when defining priority areas for the installation of more costly mains sewerage, and it is important for water service utilities to make strong representation of drinking-water quality interests with municipal government. Even where main sewerage systems are present, a potential hazard to groundwater quality can still exist as a result of major sewer leakage to the ground, and some wastewater disposal and reuse practices.

Research worldwide suggests that leakage from defective urban sewers is an important source of nutrients and microbial pathogens in groundwater. Some construction defects and the increasing age of sewer conduits make them a potential groundwater pollution source under certain shallow hydrogeological conditions or if a groundwater drain if installed below the water table (Figure A7). Groundwater inflows to deeper main sewer collectors often represent a notable increment in the total flow to wastewater treatment plants.

Wastewater treatment aims to speed up the processes by which wastewater is purified, and the higher the level of treatment the larger is the sludge residue. Wastewater treatment plant residues should be disposed of in such a way as to avoid groundwater contamination and, if land application is used, its location needs to be carefully controlled to avoid groundwater pollution.

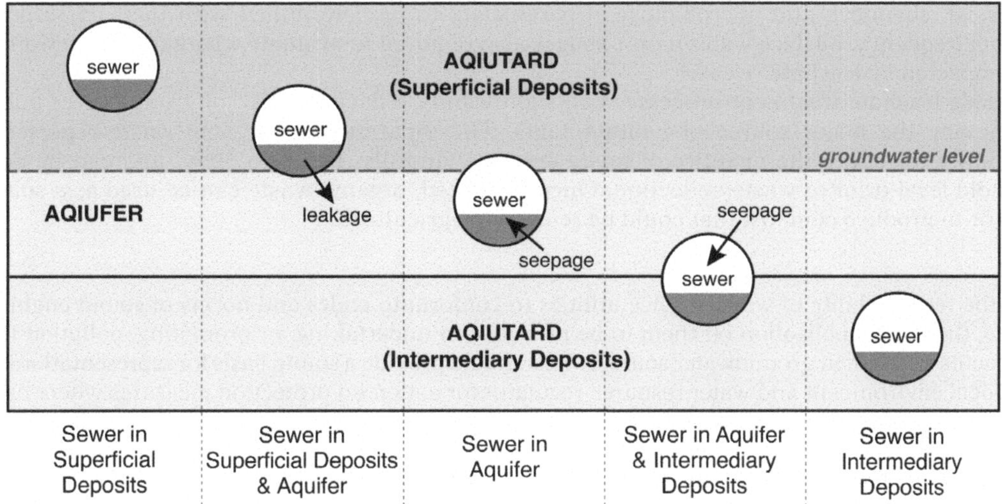

Figure A7 Hydrogeological influence over sewer leakage and seepage regimes (modified after Gogu *et al.*, 2019).

In developing countries, land application of primary and secondary treated wastewater by flood irrigation is still widely practiced as an alternative to a costly advanced wastewater treatment plant. But such approaches have to be very cautiously applied to avoid serious groundwater pollution, and the presence of certain types of industrial wastewater greatly complicate the management of the process.

In regions facing severe water resource shortages, appropriately-treated domestic wastewater could be used to recharge aquifers via infiltration ponds and/or land spreading, but very careful spatial planning and operational management is required to avoid excessive impact on groundwater quality.

Urban solid waste management

Rapid urban population growth and economic development have led to increased generation of solid wastes, making its management a major challenge for countries all over the world. In most cases, solid waste disposal into landfills is the lowest cost option, even when landfills are highly engineered. Landfill sites are commonly used for the co-disposal of various types of waste, both domestic and industrial, although the most hazardous forms of industrial waste require a separate approach.

Some of the wastes disposed into landfill generate leachate, which may contain various heavy metals (such as Cd^{2+}, Cr^{3+}, Cu^{2+}, Pb^{2+}, Ni^{2+} and Zn^{2+}), xenobiotic and aromatic hydrocarbons, phenols and microplastics which pose a serious hazard to the environment, to the underlying groundwater and to human health. This hazard is greater where no control over landfill disposal exists, and robust control measures can ensure that certain waste fractions are not disposed of at inappropriate locations.

Leachate composition is variable, and its volume varies widely with the waste moisture content and whether atmospheric precipitation or surface drainage reach the body of the landfill. Leachates move through the waste layers of a landfill transporting numerous pollutants towards the soil and underlying groundwater.

Various structural characteristics require consideration when assessing the potential impact of landfills on groundwater, notably the presence of an impermeable base layer and drain for leachate collection and safe disposal. Landfills themselves should be monitored for both liquid and gaseous contaminants.

Where landfills are located immediately above aquifers, a local defensive groundwater monitoring plan needs to be in place. Groundwater contamination is typically detected by using monitoring wells, both up-flow and down-flow of the landfill itself. A monitoring programme should define a list of physical, chemical, and microbiological parameters to be determined with their corresponding sampling frequency. Surface water monitoring is also required to evaluate whether surface waters are being impacted by leachate releases.

Reliable leachate treatment processes can significantly reduce the risk of groundwater pollution by removing the major source of contaminants. The separation of organic waste represents an opportunity to reduce the quantity of waste entering landfills (by up to 50%), and can be done at household level prior to waste collection. Once separated, organic waste can be used as a source of biogas or to produce compost that could be re-used in agriculture.

Responsibility for groundwater protection

Given the responsibility of water service utilities to conform to codes and norms of sound engineering practice, there is an obligation on them to be proactive in undertaking, or promoting, pollution hazard assessments for all their groundwater sources. This should provide a sound basis for representations made to the local environment and water resource regulator for action on protection measures where needed.

The initiative on groundwater source protection zones must normally be taken by the water utility, but undertaken in close coordination with the water resource agency. Even where no adequate pollution control legislation or agency exists, it will normally be possible for the local government or municipal authority to take protective action under decree in the greater interest.

The procedures for groundwater pollution hazard assessment presented also constitute an effective vehicle for initiating the involvement of relevant stakeholders (including water user interests and potential groundwater polluters). The proposed assessment procedure require participation of a number of qualified professionals – notably a groundwater specialist /hydrogeologist and an environment engineer/scientist – normally supported by auxiliary staff with a local office base and field transport.

Although the methodology presented is relatively simple, it will be necessary for the professional staff involved to have a reasonable understanding of groundwater pollution. Moreover, skills will need to be developed (both on the job and through consultation) in ranking some of the more subjective components of aquifer pollution vulnerability and subsurface contaminant load assessment.

The boundaries of an assessment area must be defined on a physical basis to include an entire aquifer or groundwater sub-catchment within an aquifer, so as always to include the probable recharge area of the system under consideration. It should be possible for an appropriate team to complete a groundwater resource and supply pollution hazard assessment within 3–18 months, depending on the size and complexity of the area under consideration, and the modelling strategy adopted.

How can groundwater monitoring be strengthened for adaptive management?

General philosophy

Groundwater quality monitoring is an important element of adaptive groundwater management, and in particular quality protection (Tuinhof et al., 2006). Monitoring protocols should plan to track groundwater quality status and trends, and how they respond to management actions. They thus should provide key information on the effectiveness of existing management actions and the need for iterative adjustments to secure the desired quality objectives.

The cycle of 'management action' followed by 'monitoring response' and 'policy adjustment' is then repeated over periods of years, incorporating new information from the monitoring network and predictive modelling.

A water utility's groundwater monitoring programme should be an integral part of their **Water Safety Plan (WSP)**. The monitoring programme should be based on existing legal provisions (directives, guidelines and regulations), and take into account the specifics of each groundwater source. Water quality standards (threshold values) need to be established at the appropriate level, and it is important to ensure that the management measures employed comply with operational limits.

Groundwater monitoring should include testing for specific water quality determinands and their associated indicators, selected according to the local potentially-polluting activities identified. In the case of groundwater, routine quality monitoring must be supplemented by inspection of the potentially-polluting activities and the integrity of sanitary measures in the source capture area.

Professional competence

Developing a sound understanding of groundwater pollution hazards requires inputs from a team of professionals covering the fields of hydrogeology, sanitary engineering and environmental science – and will involve investigation in institutions and agencies beyond the water utility. But water utilities should take the initiative to establish such interdisciplinary teams, and play the leading role in promoting effective groundwater source protection through cooperation with other stakeholders. Such supporting programmes are essential to the delivery of safe drinking water and in parallel require:

- the appointment of a senior staff member (such as the water safety manager) to be responsible for the entire programme
- establishment of an internal communication strategy to ensure information from monitoring is acted upon promptly
- adoption of codes of good practice for groundwater quality issues and laboratory analyses
- implementation of staff training and awareness programmes
- regular information exchange with the water resource regulator and other stakeholders as appropriate
- definition of a risk communication strategy to the public, to be implemented at times of elevated risk
- securing stakeholder commitment and collaboration contracts to promote groundwater protection.

Monitoring objectives

Groundwater quality monitoring can have a number of objectives:

- building a picture of expected raw water quality and its variability
- detecting a decline in raw water quality and its associated risks
- water intake management and, where necessary, designing a water treatment plant
- supporting a source management plan, as part of risk assessment processes, and assessing effectiveness of management measures.

The WHO Guidelines define two modes of monitoring:

- **operational monitoring** to identify – as the first line of protection of human health – whether water is likely to comply with potable water quality parameters or if groundwater abstraction is mobilising anthropogenic pollutants or geogenic contaminants, and also to monitor the effectiveness of any pollution control measures implemented.
- **audit monitoring** to determine whether all potable water quality parameters are being complied with in a given source – as the second line of protection for human health.

The operational and audit monitoring results for a given groundwater source will indicate whether there is a need for further investigative monitoring, to evaluate a specific threat to groundwater quality in detail.

Monitoring network design

A key process of groundwater monitoring programmes is network design – deciding where, what and how often to observe. This will depend on:

- the monitoring objectives
- the desired reliability of statistical data obtained
- the type and size of the groundwater body under observation
- the type of groundwater pollution anticipated.

The initial network design should include all existing waterwells and consider drilling some dedicated piezometers. While using only existing wells may initially seem tempting due to the associated low cost and rapid implementation, it has a number of disadvantages. Existing wells with long screen intake lengths tend to mix groundwater of widely different genesis, whereas dedicated piezometers can be carefully located in space and depth according to the conceptual hydrogeological model and the specific objectives of the monitoring programme.

Different groundwater monitoring strategies can be adopted (Figure A8):
- **offensive detection monitoring** around potential sources of pollution with analytical parameters selected specifically according to the type of pollution source, where the goal is early detection of groundwater pollution from known potential sources – this approach is expensive and must be used selectively
- **defensive detection monitoring** of potable groundwater sources to provide early warning of potential pollution – this requires sound understanding of the local groundwater flow system and contaminant transport routes, especially with respect to the depth selection for monitoring piezometers
- **site assessment of known contamination,** which is similar to offensive monitoring and serves two possible purposes – to confirm the efficiency of natural contaminant attenuation or the effectiveness of engineering remediation measures to contain pollution.

When designing groundwater monitoring networks, a number of issues need to be taken into account:
- the hydrogeological structure of the catchment area
- the hydrochemical conditions in terms of vertical concentration gradients and redox variations (Figure A9)
- the groundwater flow conditions and vertical hydraulic gradients
- land use in the catchment area and assessment of the pollution hazard potential of urbanisation, agricultural practices and industrial activity.

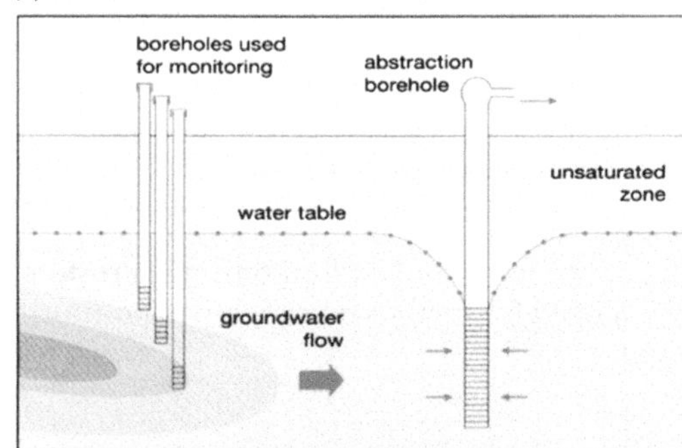

Figure A8 Groundwater quality monitoring networks for specific management objectives.

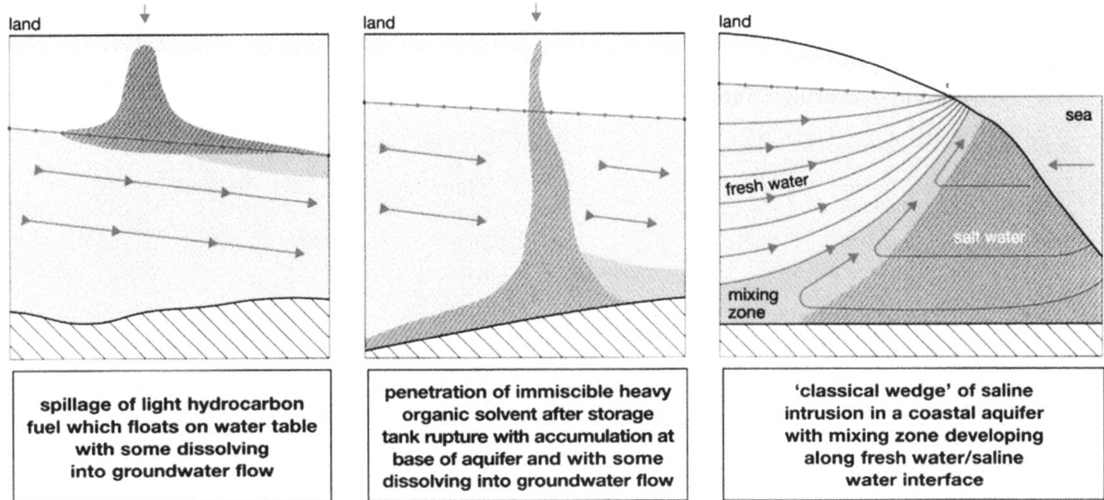

Figure A9 Processes causing major vertical variation in groundwater quality which need to be detected by monitoring networks.

Table A9 Selected parameters for possible groundwater quality monitoring.

(a) Pathogenic Organisms		
Pathogenic Species	**Persistence in Water**	**Chlorine Resistance**
Protozoa		
Cryptosporidium parvum	Long	High
Giardia intestinalis	Moderate	High
Entamoeba histolytica	Moderate	High
Bacteria		
Campylobacter jejuni	Moderate	Low
Escherichia coli	Moderate	Low
Leptospira spp	Long	Low
Salmonella typhi	Moderate	Low
Shigella spp	Short	Low
Vibrio cholerae	Short-long	Low
Viruses		
Enteroviruses	Long	Moderate
Hepatitis A & E	Long	Moderate
Noroviruses	Long	Moderate
Rotaviruses	Long	Moderate

(Continued)

Table A9 Selected parameters for possible groundwater quality monitoring (*Continued*)

(b) Chemical Compounds with WHO Drinking Water Guidelines	
Naturally-Occuring Contaminants	
Arsenic	10
Barium	700
Boron	2400
Fluoride	1500[b]
Selenium	40
Uranium	30
Agricultural Pollutants	
Nitrate	50 000[c]
Nitrite	3000[c]
Alachlor	20
Aldicarb	10[d]
Atrazine	100[e]
Carbofuran	7
Chlordane	0.2
Chlorotoluron	30
2,4D	30
Dichlorprop	100
Fenoprop	9
Isoproturon	9
Lindane	2
MCPA	2
Mecoprop	10
Methoxychlor	20
Metolachlor	10
Simazine	2
Industrial & Community Pollutants	
Cadmium	3
Chromium	50[g]
Mercury	6[h]
Benzene	10
Carbon tetrachloride	4
Dichloromethane	20
Ethylbenzene	300
Pentachlorophenol	9
Tetrachloroethene	40
Toluene	700
Trichloroethene	20
Xylenes	500

[a] expressed as μg/l (ppb), although some determinands are usually given in mg/l (value in ug/l is divided by 1000 to report in mg/l) – note some countries have lower and/or higher values for certain contaminants
[b] but consider all intake
[c] can also be wastewater derived
[d] both sulfoxide and sulfone
[e] hydroxyatrazine 200 μg/l
[f] now mainly non-agricultural use
[g] as total chromium
[h] as inorganic mercury

Selection of monitoring parameters

The selection of monitoring parameters should relate to their reliability and sensitivity in detecting incipient groundwater quality degradation of the suspected type. A full range of potential parameters is given in Table A9. The frequency of monitoring will depend on the dynamics of the groundwater system and nature of the pollution threat.

In some groundwater systems, monitoring may need to become more targeted during times of known elevated risk, but in general (given the large storage and slow reaction times of most aquifers) relatively infrequent monitoring (every 3–6 months or more) is acceptable.

Chemical indicators can be deployed in groundwater monitoring to reduce the analytical workload, providing that the indicator(s) chosen are readily measurable (especially *in situ*) and respond reliably to the type of pollution suspected.

The establishment of a groundwater quality monitoring programme requires that the following elements are carefully considered:
- formulation of working methodologies for collecting information
- identification of suitable analytical techniques (Table A10)
- identification of appropriate analytical equipment
- definition of analytical quality control procedures for both laboratory and field tests.

It is important to perform a financial analysis of the long-term groundwater monitoring plan that considers the number and location of sampling sources that will be included in each phase of the work, the required analytical equipment and the transport costs. The available budget will have to be

Table A10 Sampling procedures and warnings for groundwater quality parameters.

Determinand Group	Sampling Procedure	Preferred Materials	Storage Time/ Temperature	Operational Difficulty/Cost
Major Ions Cl, SO_4, F, Na, K	• 0.45 μm filter only • no acidification	any	7 days/4°C	minimal
Trace Metals Fe, Mn, As, Cu, Zn, Pb, Cr, Cd, and so on.	• sealed 0.45 μm filter • acidify (pH <2) • avoid aeration through splashing/head space	plastic	150 days	moderate
N Species NO_3, NH_4 (NO_2)	• sealed 0.45 μm filter	any	1 day/4°C	moderate/low
Microbiological TC, FC, FS	• sterile conditions • unfiltered sample • on-site analysis preferred	dark glass	6 hours/4°C	moderate/low
Carbonate Equilibria pH, HCO_3, Ca, Mg	• unfiltered well-sealed sample • on-site analysis (pH, HCO_3) (Ca/Mg at base laboratory on acidified sample)	any	1 hour (150 days)	moderate
Oxygen status pE(EH), DO, T	• on site in measuring cell • avoid aeration • unfiltered	any	0.1 hour	high/moderate
Organics TOC, VOC, HC, CIHC, and so on.	• unfiltered sample • avoid volatilization • (direct absorption in cartridges preferred)	dark glass or teflon	1–7 days (indefinite for cartridges)	high

taken into consideration in the design of the sampling programme, and additional financial resources sought if necessary.

Groundwater quality monitoring programmes are also conceived for other reasons, such as:

- to obtain urgent answers to pressing concerns
- as a formal duty of accountability for water quality
- as part of a broader research and development activity.

The process of obtaining reliable monitoring information can be represented as a spiral which shows its dynamic interaction with decision making in groundwater management and protection activities.

doi: 10.2166/9781789063110_0023

Part B

Groundwater Hazards for Urban Infrastructure

B1 CHARACTERISATION OF PROBLEMS

Why is groundwater important to city infrastructure planning?

The impact of groundwater on subsurface infrastructure has become an important subject due to the continuous expansion of modern cities. As cities expand downward, underground spaces replace urban surfaces, especially for utility services (cables, sewerage, drainage), transportation routes (subways, tunnels, passages), storage facilities (warehouses, cellars, parking spaces), and other diverse uses. These infrastructure elements and their relation with groundwater have to be carefully considered in urban planning (Figure B1).

City planners and engineers need to have an understanding of groundwater. To provide this understanding, water resource agencies, water utility companies and geological surveys need robust datasets on groundwater at city scale (Figure B2). This data should be made accessible for subsurface planning, in datasets appropriate to different scales of interest and different planning stages.

When studying city infrastructure plans (such as the development of a subway line or underground parking facility) it is relatively easy to identify those elements which might be significantly impacted by groundwater but have not been adequately evaluated.

Many cities are facing serious consequences as a result of a lack of detailed knowledge of the urban subsurface environment. Current practice in the development of underground infrastructure widely involves only local site investigation of groundwater conditions but fails to reach understanding of the dynamics of the urban groundwater system. This systematic mistake is usually due to the narrow perspective of civil engineering companies and infrastructure developers, and often goes unchallenged by municipal government departments.

The consequences of inadequate consideration of subsurface conditions in urban infrastructure development can be far reaching in economic, social and environmental terms. Poor appreciation of ground conditions, and especially failure to consider the dynamic status of groundwater occurrence, is recognised as the largest cause of construction project delay and cost overrun.

In the urban environment, quantifying and modelling groundwater flow is a demanding task due to the general lack of data availability and the physical separation of municipal government staff from hydrogeological specialists. The urbanisation process impacts groundwater, and groundwater can impact the subsurface infrastructure. Since urbanisation is related to sustainable development, 'green

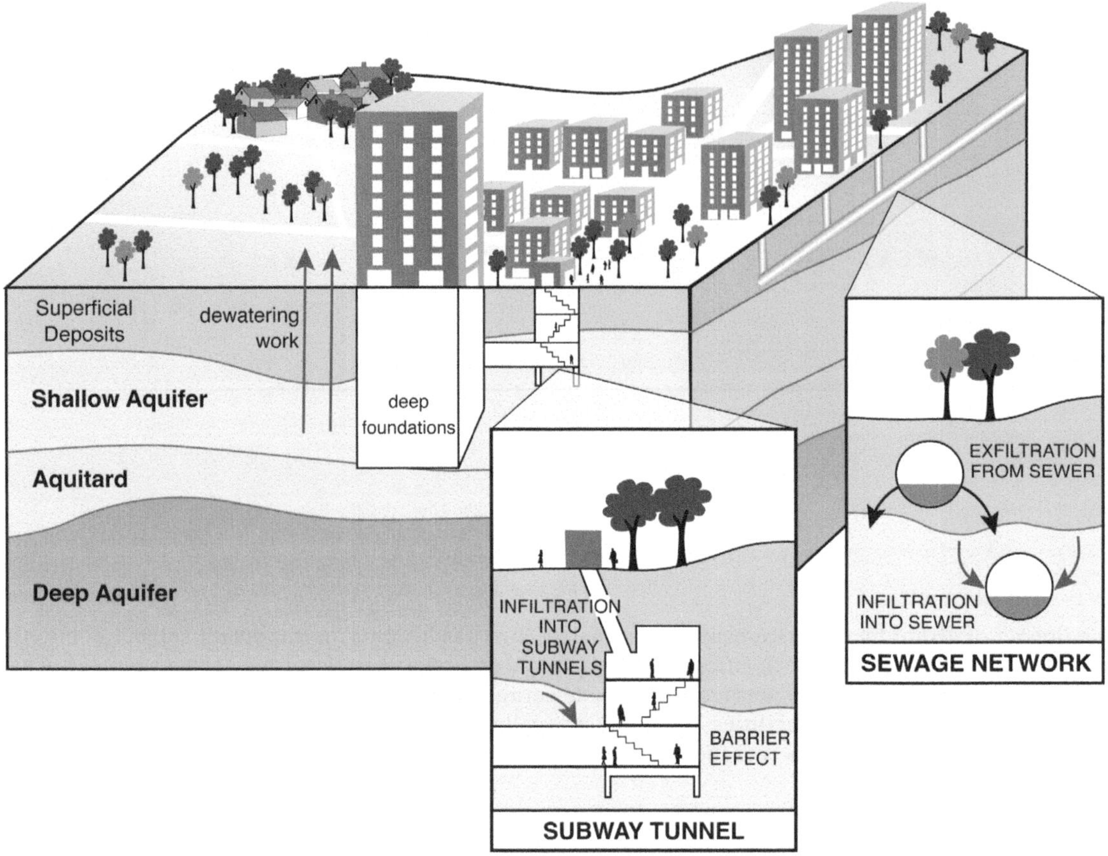

Figure B1 Interactions between urban subsurface infrastructures and groundwater.

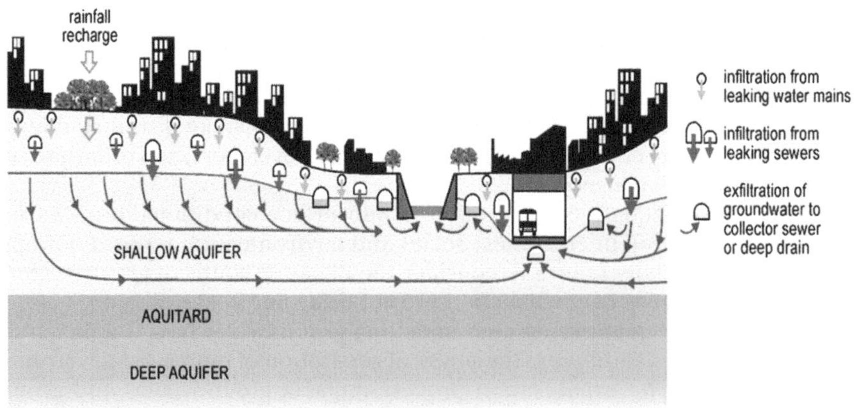

Figure B2 Typical relation between shallow groundwater system and urban infrastructure.

infrastructure planning' is the single reliable solution. Future project design and implementation should refer to urban groundwater studies, and consider many nature-based solutions which are linked to groundwater.

How does urbanisation impact the groundwater regime?

Urbanisation changes the shallow groundwater regime (Table B1) by:

- modifying recharge processes – as a result of the extension of the impermeabilised land surface, the exfiltration from water mains and sewers (Figure B2), and the over-irrigation of gardens and parkland – usually with a net increase in local recharge
- sustainable urban drainage systems (SUDs) routing all roof and open space drainage to soakaways to augment recharge
- draining groundwater systems through the construction of main-collector sewers and some deep underground structures (railway tunnels, deep basements)
- installation of groundwater 'flow barriers' through the construction of metro stations, road tunnels and deep foundations
- substantially increasing subsurface contaminant load as a result of seepage from sewers and *in situ* sanitation, inadequate storage and/or disposal of liquid and solid wastes, and accidental chemical and oil spills at the land surface.

These changes in recharge and discharge usually result in significant changes in groundwater level (hydraulic head). Both rising water table (causing flooding risk to basements, inflow to sewers and ground liquefaction risks) and falling water table (generating land subsidence and building foundations) are both possible.

What data are needed to diagnose subsurface management needs?

City-scale groundwater monitoring and subsurface investigation constitute important components of sustainable urban development. The impact of future underground structures can be systematically assessed and potentially costly hazards avoided. The complexity of the groundwater system and intensity of use of subsurface space will determine the data density required to achieve a robust understanding of urban groundwater, and will vary widely from city to city.

The development of a city-scale hydrogeological model constitutes an important milestone for future work on subsurface infrastructure issues. It allows various scenarios to be simulated and, within this framework, interconnected hydrogeological and geotechnical studies related to urban

Table B1 Classification of subsurface structures and their potential effects on groundwater.

Subsurface Structure	Potential Impact	Consequences	
		Groundwater System	Subsurface Structure
Road and Railway Tunnels	• drain effect; • barrier effect	• discharge from system • variation of hydraulic head	• uplift or inflow
Basements and Foundations	• drain effect; • barrier effect	• discharge from system • variation of hydraulic head	• inflow/flooding
Water-Supply Network	• recharge effect	• increased hydraulic head	• loss of water-supply capacity/revenue
Sewer Network	• recharge effect; • drain effect	• infiltration and pollution • increased hydraulic head	• increased wastewater flow to treatment plant

subsurface infrastructure can be implemented. Ultimately, such a model represents a powerful tool for urban groundwater management.

Who has institutional responsibility for urban groundwater resources?

Groundwater in and around urban areas, and the processes affecting it, is generally addressed by procedures and regulations instructed and operated by a range of different institutions:

* permits for the construction and operation of waterwells are generally issued by the water resource agency
* the production and distribution of public water-supply is organised by the water service utility
* urbanisation and subsurface construction and drainage is usually the responsibility of municipal government
* the development of sewerage systems is usually undertaken by the water service utility but to standards dictated by the environmental or public health authority, as is the case for industrial effluents and other waste.

The level of available data and quality of monitoring systems is very much a function of the scientific awareness and logistic capability of the municipal government agencies, water resource agency and the water service utility, and can vary from completely adequate to totally inadequate. Various factors are often quoted as being responsible for impeding municipal government involvement in urban groundwater (Table B2), but these need to be resisted.

Many cities face the consequences of the unavailability of adequate understanding of the interaction between groundwater and the urban subsurface. Current practice in the development of subsurface infrastructure is too widely reliant on a narrow geotechnical investigation for the specific new structure in question, which fails to grasp (and design for) the dynamic nature of the local groundwater system.

Can your agency contribute effectively to urban groundwater management?

Each institution makes managerial decisions in the fields of activity which it commands. However, these decisions need to be coordinated within the broader institutional framework, such that all important topics get covered. In order to achieve an acceptable level of management of groundwater resources in the urban environment, it is necessary to have a large volume of data from different domains, which implies the need for involvement of a corresponding group of institutions. It is in this context that the potential involvement of municipal authorities in understanding urban groundwater needs to be considered.

Table B2 Factors impeding municipal government involvement in urban groundwater.

Factor	Outcome
Assumption or perception (sometimes reinforced by legislation) that groundwater management is sole responsibility of another organisation	Responsibility entrusted entirely to environment or water resources agency or basin authority
Understanding of potential interaction between groundwater and urban infrastructure missing	Subsurface structures developed with only geotechnical site investigation
Density and frequency of groundwater data for robust understanding is absent	Large investment supported by municipal authority required
Municipality size too small to allow effective contribution to groundwater management	Insufficient authority over required land area to be effective

In this context it is important to consider the municipal government's potential contribution to the monitoring of groundwater systems, the surveillance of potential pollution sources and the effects of groundwater drainage by subsurface infrastructure elements.

B2 ESSENTIAL STEPS FOR INTEGRATED ACTION

To improve urban development, more effective use of subsurface space is essential. Most municipal governments will benefit considerably from drawing up an 'integrated action plan' for the managed development of their subsurface space, which must include a number of components:
- procedures to bring groundwater considerations into urban planning
- knowledge needed for effective subsurface urban planning
- practical administration of urban subsurface development.

Which procedures can ensure that groundwater considerations are always included in urban planning?

Improved interaction and communication between urban planners and geoscientists (in appropriate agencies) needs to be fostered to ensure that groundwater considerations are taken fully into account in subsurface development projects. In particular, geoscientists need to better appreciate how and when different types of information can enter into both the strategic and detailed urban planning hierarchy. Information about the subsurface needs to be technically reliable, well organised, regularly updated and easily integrated into the planning and construction process. The type and amount of information will vary with the planning level and the planning tasks.

Guidelines should be produced to support planning, and 'specific city requirements' identified in urban planning procedures. These will differ between regions, countries and municipalities, but are shaped by:
- the policy framework (legislation, directives and agreements)
- the planning scale, phase and stage
- specific features of the city (economy and environment).

A 'city analysis' should to be based on the local subsurface position, in terms of hydrogeologic conditions and infrastructure development and/or the resources of the city, and adapted to cover expected urban development and available subsurface information.

A key procedure to help ensure that groundwater considerations are included in urban planning is to have an up-to-date detailed urban hydrogeological map or GIS system available. This geospatial database and conceptual model for the groundwater system, potentially interacting with subsurface infrastructure, will require the information listed in Table B3, and will require systematic updating at 5-yearly intervals. The model should capture the interaction between shallow groundwater and the urban infrastructure, and include data on the natural controls on groundwater level (such as precipitation and surface water features), and the impact on groundwater levels of the existing urban infrastructure (impermeable surfaces, waterwell abstractions, dewatering systems, water service conduits, transportation tunnels, etc.). Such models must be based primarily on data from a well-developed groundwater monitoring system.

The knowledge development process for urban groundwater usually begins with the use of existing data, and continues with additional data capture, which is followed by knowledge building in conceptual hydrogeological models (Figure B3). These models can then be used to describe processes and simulate scenarios interactively with urban planners and infrastructure engineers. The decision-making process can then be optimised on the basis of these subsurface models, and professional interaction will then encourage a detailed monitoring of shallow groundwater and the hydraulic interaction between subsurface infrastructure elements.

Table B3 Data requirements for a geospatial database.

Element	Data Requirements
Groundwater System	• identification of aquifers, aquitards and aquicludes • age of the related geological deposits • their lithological and granulometric characteristics • extent and thickness of the aquifer layers • hydraulic properties (conductivity, transmissivity) • areas of groundwater recharge/discharge and flow direction • groundwater level trends from time series • relationship with the surface hydrological features • land use map of the recharge areas • groundwater quality distribution • aquifer pollution vulnerability maps • hydrogeological model of groundwater flow
Groundwater Abstraction	• well inventory, including owner and water use • well drilling depths, screen positions, pump capacities • inventory of closed and abandoned wells
Groundwater Pollution Threats	• existing and potential point or diffuse pollution sources • types of potentially polluting compounds • potential routes of penetration into groundwater
Groundwater Management	• name of competent institution or authority • institution/authority's field and decisional hierarchy details
Subsurface Infrastructure	• maps of water service distribution networks • maps of sewer network and areas with *in situ* sanitation • depth of water mains and sewer installation • any information available relating to network gains and losses • major building foundation depths • inventory of construction dewatering work

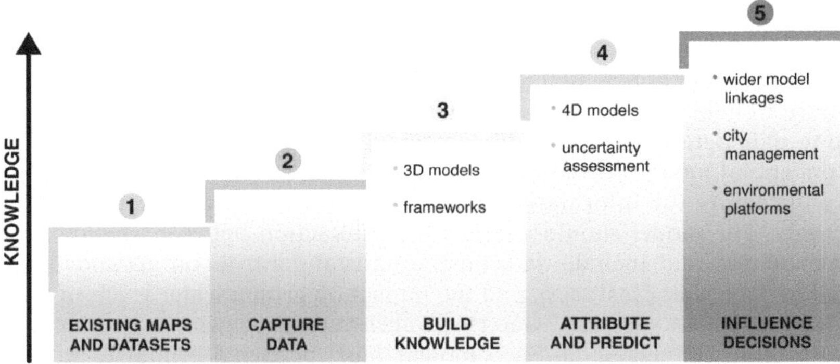

Figure B3 Key information levels in the urban groundwater knowledge building process.

What types of groundwater information are essential for the urban planning process?
Reliable conceptual models of urban groundwater can only be developed after establishing a groundwater balance, which includes water-supply mains losses (often a major urban groundwater recharge component), and sewerage network gains and losses. It is also necessary to understand whether hydraulic barriers in the saturated aquifer have been introduced by the construction of transportation tunnels, deep building foundations and underground car parks.

The shallow urban geological conceptual model is used as the framework for the hydrogeological model and the hydrogeological units comprising the shallow aquifer system. Well pumping tests and aquifer grain-size analyses have to be performed to characterise these units. It should be possible to have the geometric parameters needed to identify any groundwater flow barriers produced by the presence of subsurface engineering works and to assess the probable hydraulic behaviour of water-supply mains and sewerage networks.

Shallow groundwater monitoring is a cyclic process, in which dedicated networks will initially be required for each specific site objective. City-scale groundwater monitoring networks can then be developed from the dedicated site networks, and should include mainly shallow, but also a few deep, observation wells.

Urban planning processes differ significantly around the world, but there are common priorities as regards the types of datasets and technologies used to enhance our understanding of the urban subsurface: Priority datasets include:
- geotechnical properties of strata
- groundwater occurrence and surface water interaction
- location of subsurface buildings and tunnels
- location of water service pipes
- contaminated land distribution and history.

Most cities have a huge amount of geotechnical data on their underground space. For subsurface construction projects, geotechnical engineering standards (such as EUROCODE or ASTM's) have to be followed to investigate and specify the physical and mechanical properties of ground materials, including soils, aquifers and bedrock. This geotechnical data allows engineering organisations and construction companies to establish the physical characteristics of the ground to ensure safe construction and avoid geotechnical hazards.

How should the development of urban subsurface space be administered with regard to groundwater considerations?

The best way of administrating the subsurface space of a city is for the urban planning authority as regards groundwater conditions is to provide development companies and construction contractors with a set procedure to follow in relation to groundwater assessment (Table B4).

From the planning perspective it is also very important to make it clear to developers which hydrogeological characteristics potentially pose serious constraints on certain types of subsurface development. To aid this process the urban municipal authority should:
- task and contract an institution to maintain an up-dated hydrogeological model and information system on open file for the use of urban subsurface developers

Table B4 Recommended procedure for construction manager on reporting groundwater data for a specific subsurface site.

Reporting Procedure
• Elaboration of a geotechnical study of the construction site (using EUROCODE or similar), including hydrogeological conditions from latest urban hydrogeological map
• Monitoring of site pre-construction to assess the potential impact of the new structure on the groundwater regime (water levels and quality)
• Submission of interim groundwater study to urban planning authority for approval, with remedial solutions proposed to mitigate any unacceptable impacts
• Final groundwater study post-construction, with measurement of the specific impacts caused by installation of the new structure
• Submission of final groundwater study to urban planning authority, including description of any disturbance to local groundwater regime, for incorporation in urban database

- establish urban planning targets that specify constraints on the level of anthropogenic impact on groundwater levels and quality resulting from urban subsurface construction
- promote collaboration and data exchange between urban subsurface developers and the municipal authority and its supporting institution.

A periodic groundwater status report is an effective instrument for building awareness of the potential constraints and challenges of hydrogeological conditions for subsurface construction. A sound urban shallow groundwater monitoring network will provide essential data for the management of subsurface space, and this needs to be operated efficiently by a specialist organisation to a plan which specifies the variables to be monitored and the financial arrangements for the work.

The operation of a public groundwater observation well network helps to make groundwater 'more visible'. The monitoring data will need to be posted on the internet, and the web-platform will require periodic update. A good example is the practice of several Dutch municipalities which already operate public groundwater monitoring websites (https://maps.waternet.nl/kaarten/peilbuizen).

doi: 10.2166/9781789063110_0031

Part C

Examples of Urban Groundwater Management

C1 BANGKOK (THAILAND)
Major reduction and redistribution of groundwater abstraction to control land subsidence (data from Buapeng & Foster, 2008)

Greater Bangkok has developed to now occupy the lower part from the Chao Phrayh Basin, which is underlain by 500 m of interbedded alluvial and marine sediments from the Pliocene-Pleistocene age, containing eight semi-confined 'aquifer horizons' (recharged from the north) and overlain by a confining Holocene clay. Widespread exploitation of groundwater from the 2nd, 3rd and 4th sub-aquifers beginning in the 1950s, mainly for the Metropolitan Waterworks Authority (MWA) water-supply, reached a level of about 500 Ml/d (500 million l/d) by 1985, and caused a lowering of the groundwater levels over much of the urban area to 40 m below sea-level, with evidence of significant land subsidence (Figure C1).

The initial approach to reducing groundwater abstraction was to require the MWA to close its wells in favour of development of distant surface water sources with increased water tariffs, but this triggered a massive increase in private well drilling. Abstraction reached over 2000 Ml/d by the late 1990s, with a further 400 Ml/d abstraction by three Provincial Waterworks Authorities.

The Department of Groundwater Resources (DGR) had been given powers to manage groundwater in 1977, but in 1983 these were strengthened to enable groundwater abstraction to be reduced given widespread evidence of serious land subsidence. The DGR identified 'critical areas' where well drilling must be banned, assigning powers to seal wells in areas with mains water-supply coverage, and licensing and charging for groundwater abstraction. In 1985 charges were introduced at only a nominal rate but subsequently raised under two separate components (a 'groundwater use fee' and a 'groundwater conservation fee' each reaching US$0.21/$m^3$ by 2004), with more aggressive application of sanctions and well sealing. These measures had the positive outcome of controlling groundwater abstraction and reducing land subsidence. There are now just over 4000 licensed wells operated by around 3000 owners, abstracting about 1600 Ml/d (representing 15% of the total water-supply). There has been conflict in some districts where the mains water-supply was extended but with high charges (US$0.60/$m^3$), and these were resolved by allowing private well users to continue pumping up to 10 years (their next license renewal) and retaining their wells as a back-up supply (Figure C2).

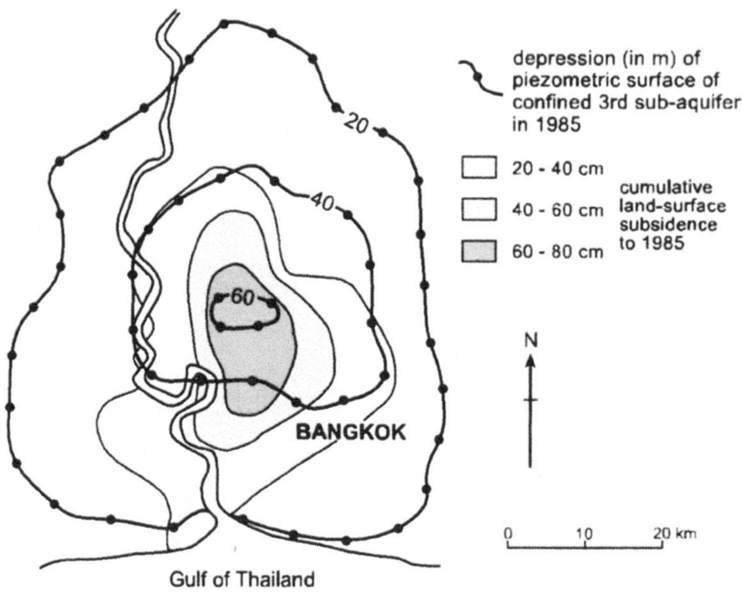

Figure C1 Distribution of land subsidence in relation to groundwater abstraction in Greater Bangkok in 1985.

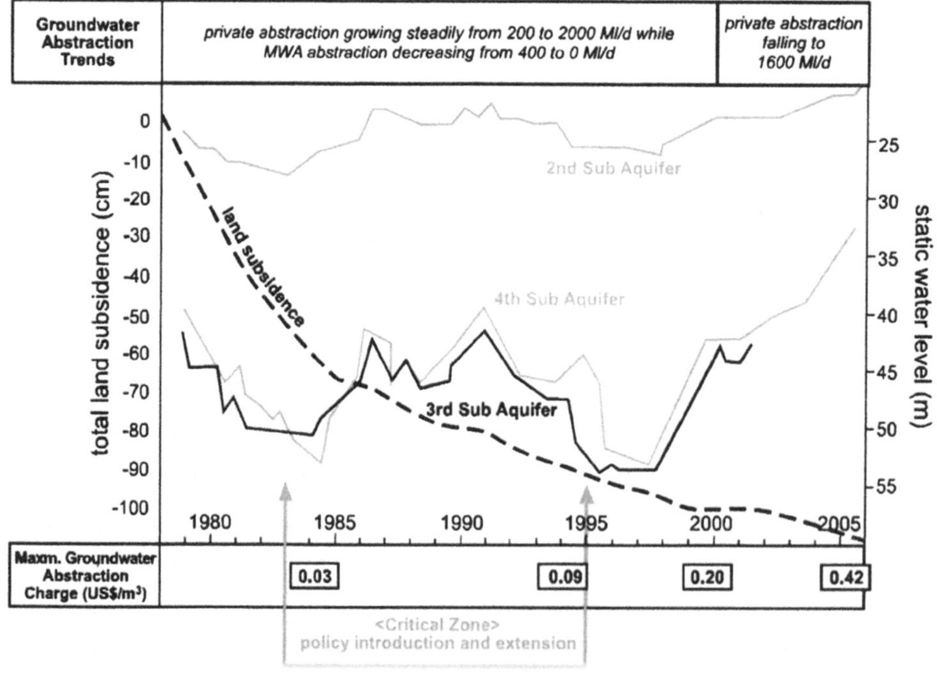

Figure C2 Timing of groundwater management interventions in Greater Bangkok with groundwater level and land subsidence response.

To undertake groundwater administration, the DGR has 25 headquarters staff and seconded three people to each of the seven provincial government offices of the metropolitan area. The Bangkok experience has demonstrated the following achieveable outcomes in groundwater management:
- ability to reverse trends in groundwater resource decline through persistent application of regulatory measures, reaching a high-level of resource fee collection
- successful targeting of groundwater management measures in objectively-defined priority areas, rather than having to apply them universally
- the capacity of a central groundwater 'apex' agency working in a decentralised fashion in unison with provincial government offices to manage groundwater
- the management and financing of long-term investments in groundwater-related environmental monitoring and research.

C2 LIMA (PERU)

Planned conjunctive use to stabilise the groundwater reserves of a critical aquifer in a hyper-arid setting (data from Foster *et al.*, 2010c)

Greater Lima extends across the Quaternary outwash fans of the Rimac and Chillon rivers occupying about 390 km² of the extremely arid coastal plain. The underlying alluvial aquifer has a saturated thickness of up to 300 m, but the upper 100 m or so (of sandy gravels) provide the best yields to waterwells. Its recharge arises from riverbed infiltration, by seepage from irrigation canals and excess irrigation to agricultural crops, parks and gardens, and by leakage from water-supply mains and wastewater infiltration, with the current total estimate put at around 190 Mm³/a (Figure C3).

During the 1960s–1980s the city grew very rapidly to a population of over 8.0 million, and water demand increased from less than 100 Ml/d in 1955 to more than 2000 Ml/d in 1997 (of which some 1580 Ml/d was provided by SEDAPAL, the municipal water-supply utility). The Atarjea Waterworks

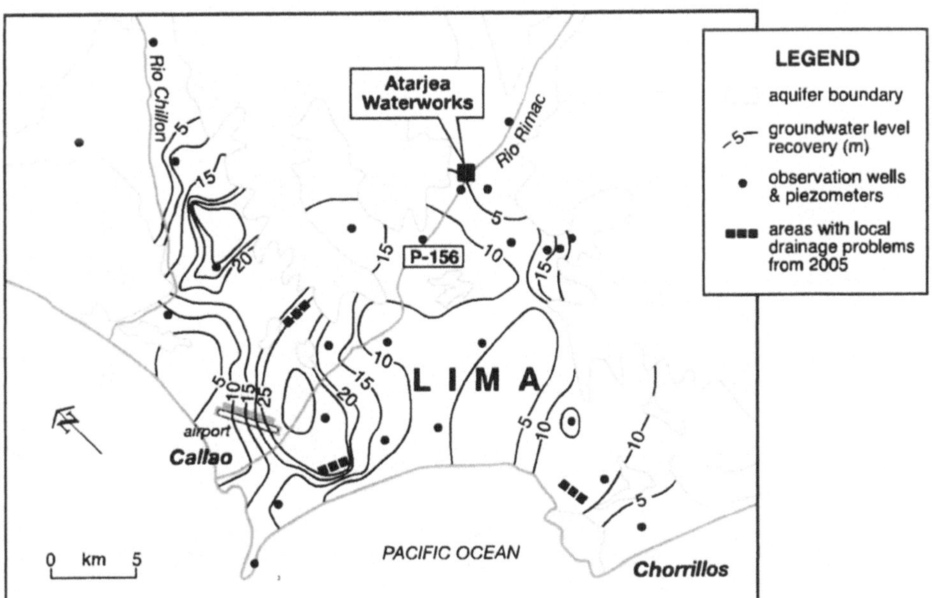

Figure C3 Distribution of Greater Lima aquifer system and its water table recovery during 1997–2003.

on the Rimac river was commissioned in 1956 and increased in capacity to 860 Ml/d in 1969, although production is impossible at times of maximum riverflow and suspended solids because of treatment problems. Of the total water-supply in 1997, 1050 Ml/d was derived from groundwater (380 SEDAPAL and about 800 industrial/commercial wells abstracting 720 Ml/d and 330 Ml/d, respectively) with a resultant continuous water-table decline of 1–5 m/a. This caused waterwell yield reductions, escalating pumping energy costs and groundwater quality decline in some areas due to saline water intrusion.

During the years 1985–95, major studies and investments were made to optimise conjunctive use of surface water and groundwater, and to manage the water demand of the city through:

- concerted micro-measurement of domestic water use (costing about US$60 million and involving metering of 700 000 users) and industrial groundwater conservation (costing about US$10 milllion), to reduce wastage
- an additional Andean surface water transfer of up to 260 Ml/d to the Rimac River (costing US$63 million), which had benefits in terms of energy generation and increased treatment plant capacity to 1500 Ml/d
- improved flexibility of water distribution in the city, to allow a much higher proportion of users to be supplied either from treated river water or from groundwater (involving installation of 66.8 km of additional water mains at a total cost of about US$22 million)
- riverbed recharge enhancement measures over 6 km of the Rimac River (costing US$12 million). including 60 transverse riverbed baffles totalling 10 000 m in length, 30 wells of 80–120 m depth with a pumping capacity of 140 Ml/d, and a comprehensive monitoring set up (Figure C4).

The institutional arrangements for the implementation of the above measures were unconventional. In 1981 a 'national supreme decree' charged and empowered SEDAPAL (the municipal water company and major groundwater abstractor) to act on behalf of national and regional government to conserve Lima's strategic groundwater reserve, given their strong 'on-the-ground operational presence'. SEDAPAL established a special office and team for the purpose, in effect acting as a contractor to government for this task, but referred all key policy decisions to IRH and SUNASS (the national water resources institute and water service regulator, respectively). The measures used included comprehensive well licensing and a ban on water welldrilling in some particularly critical areas.

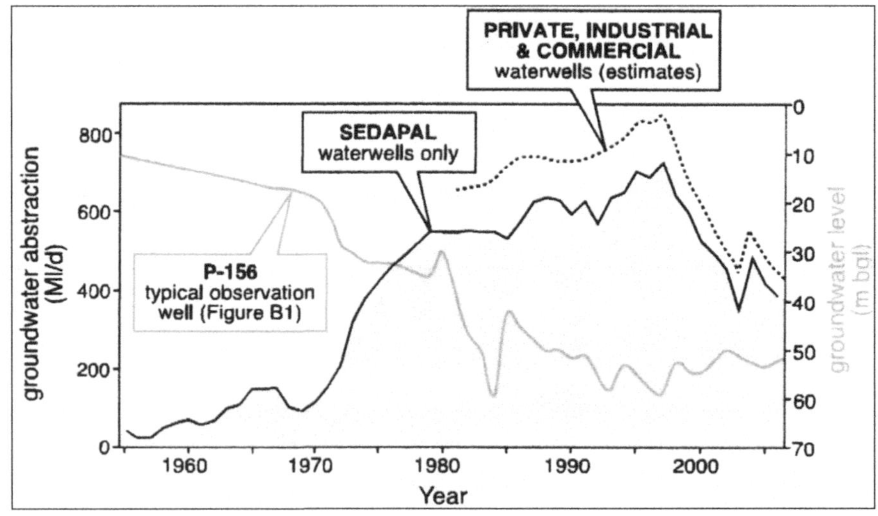

Figure C4 Historical evolution of SEDAPAL waterwell abstraction and water-table depth in the Greater Lima aquifer system.

The success of the conjunctive use and demand management measures is witnessed by the major recuperation of the water-table over wide areas by 5–30 m from 1997–2003, following a decline of 10–40 m during the preceding 10 years. Groundwater abstraction by SEDAPAL reduced from 265 Mm³/a in 1997 to a 2005–09 average of 135 Mm³/a, although capacity to reach much higher rates of production in the short term still exists. The corresponding industrial/commercial direct use figures show a decrease from 125 to 75 Mm³/a, bringing the overall proportion of urban water-supply from groundwater down to an average of 29% in recent years, from the previous level of 57%.

C3 HAMBURG (GERMANY)

Well-managed groundwater resource development with emphasis on monitoring for quality protection (data from Foster *et al.*, 2020b)

Hamburg has a population of some 2.2 million served by a municipally-owned water utility, Hamburg Wasser. In 1964, after a long transition, it switched from filtered river water to groundwater for its water-supply, Today it operates about 470 waterwells pumping some 126 Mm³/y from a shallow Quaternary alluvial aquifer and a deeper Tertiary formation (Figure C5).

Nine of the corresponding wellfield capture areas have legal status as groundwater protection zones, but three are located outside city and state jurisdiction, and their protection has to be negotiated through constructive dialogue and detailed negotiation with neighbouring authorities. In some cases conflicts have arisen over land-use practices, since the shallow aquifer is vulnerable to agricultural and industrial pollution. Groundwater supplies from the deeper aquifer are also threatened by salinisation from adjacent salt domes and require carefully-controlled pumping regimes.

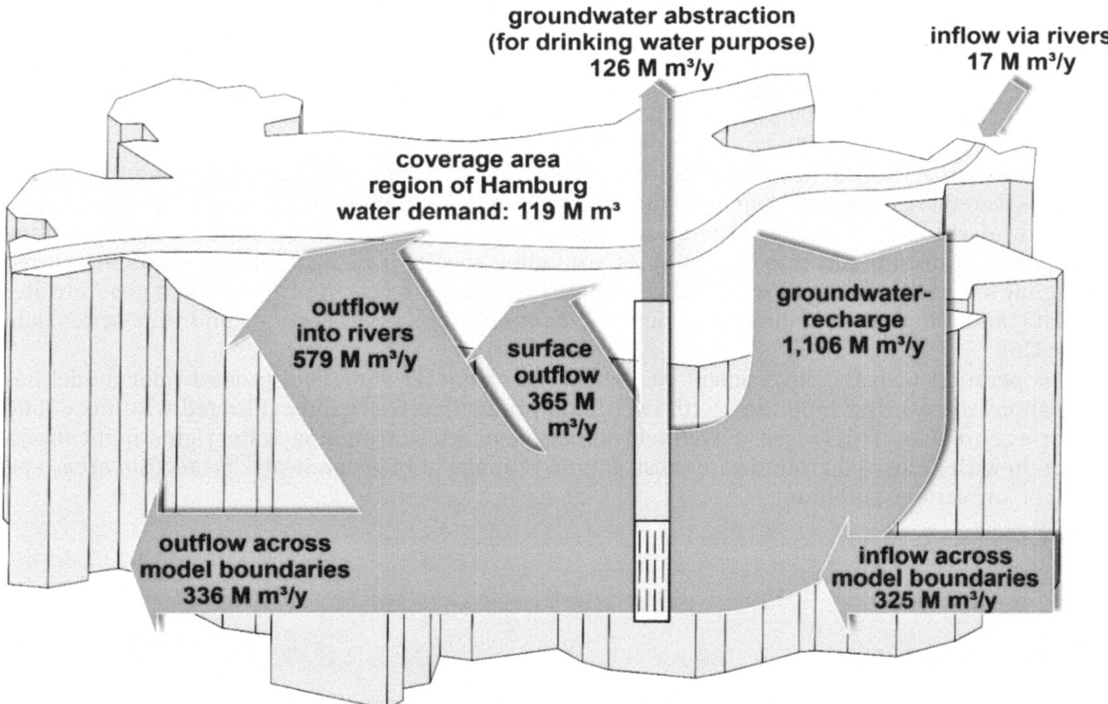

Figure C5 Groundwater balance of Hamburg area.

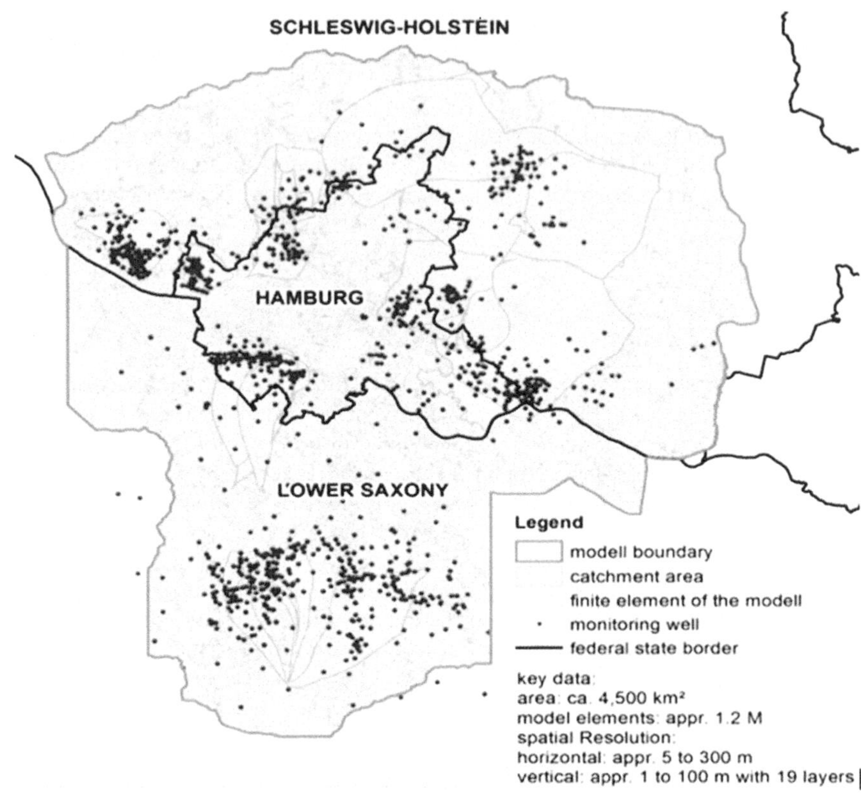

Figure C6 Configuration of Hamburg groundwater numerical model.

Since the utility is fully aware of the types of risk to groundwater quality, Hamburg Wasser maintains its own network of about 1400 monitoring boreholes which provide a full dynamic picture of groundwater flow and quality. After many years of experience, the annual groundwater sampling programme has been reduced to 350 samples, excluding special investigations. The data are stored in a digital information system which also contains hydrogeological information and groundwater level data, and allows the production of periodic contour maps, cross-sections and time-series data (Figure C6).

In cooperation with the government geological agency, a 3D numerical groundwater model has been elaborated covering 4500 km², with over 3000 production wells and calibrated with over 7000 monitoring boreholes. This is used for wellfield management decision making, water rights applications, interaction with industrial groundwater abstraction, refinement of groundwater protection areas and control of any serious pollution.

C4 BUCHAREST (ROMANIA)

Improving understanding of a shallow groundwater system to aid design and operation of subsurface infrastructure (data from Gogu, 2019)

In Bucharest, groundwater abstraction from the shallow Quaternary Colentina aquifer system – at a depth of 15–20 m below the surface, with groundwater levels of 5–10 m depth – was terminated in

2000 and a deeper aquifer is now used for urban water-supply. However, the city is faced with problems arising from interactions between shallow groundwater and the underground infrastructure including:
* 'groundwater flow barrier effects' produced by the extensively canalised Dambovita River and parallel subway tunnels of 7–23 m depth
* major groundwater inflows to trunk sewers generating an excess sewer flow of about 1 m³/s
* uncontrolled permanent or temporary dewatering systems disturbing groundwater flow and triggering subsidence in some locations.

A partnership between the urban water utility (Apa Nova/Veolia) and the Bucharest Technical University of Civil Engineering led to detailed groundwater investigations and aquifer numerical modelling being carried out, which threw light on the problem of excess sewer flows and subsurface structure interference with groundwater flow, and mobilised a broad base of urban stakeholders in data collection and monitoring. The groundwater monitoring system focuses on the first two aquifer layers and comprises 140 stations distributed in the city centre, along the subway line, along the canalised river and related artificial dammed lake, and in the outlying suburbs. This system constitutes an important milestone in the hydrogeological understanding of the city, and represents a powerful tool for groundwater management. Future urban subsurface projects can now count on accurate groundwater information (Figure C7).

The deeper Fratesti aquifer system is of the Upper-Lower Pleistocene age and represents the main formation used for water-supply in south-eastern Romania. It is spatially variable in thickness and structure, generally behaving as a confined multi-layered aquifer of up to 150 m thickness (with intervening marls and clays), but in the south of the area there is only a single aquifer layer of 40 m thickness. Excessive pumping for industrial purposes in the 1970s–80s led to a decrease in the aquifer's hydraulic head and a change in the general groundwater flow direction (Figures C8 and C9).

Currently the water utility uses the Fratesti aquifer to source an emergency drinking water-supply of 36 waterwells of 120–220 m depth, each with a capacity of 4.0–4.5 m³/s. Three specific criteria were used to locate these waterwells: yield potential, required depth and groundwater quality. Except for the sporadic presence of ammonia ($\geq$0.5 mg/l), the water quality parameters are excellent. This aquifer is also used by some hospitals and drug factories as their source of water-supply.

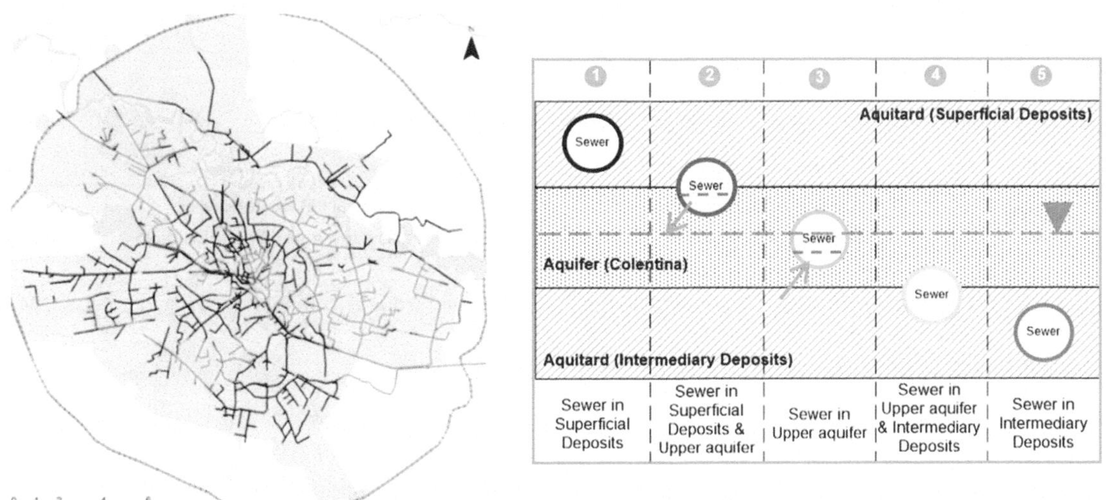

Figure C7 Urban infrastructure data used for construction of a groundwater numerical model of the shallow aquifer system.

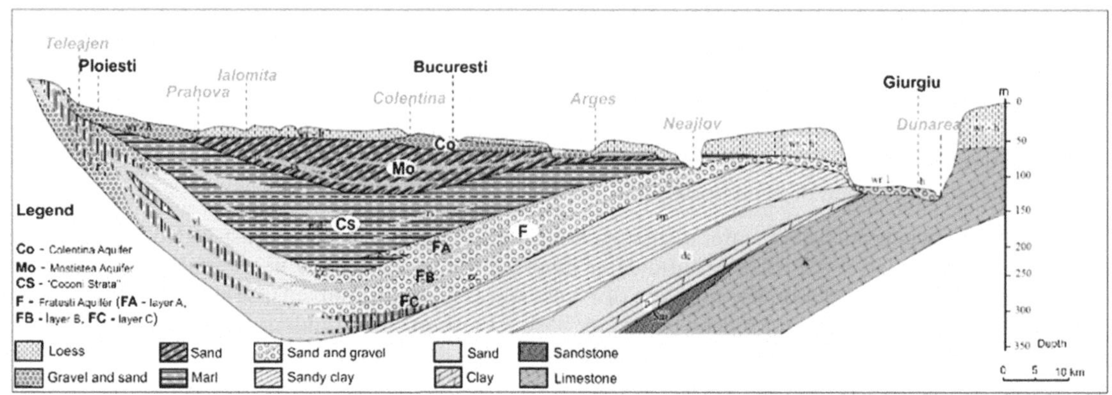

Figure C8 Hydrogeological cross-section of the Bucharest area.

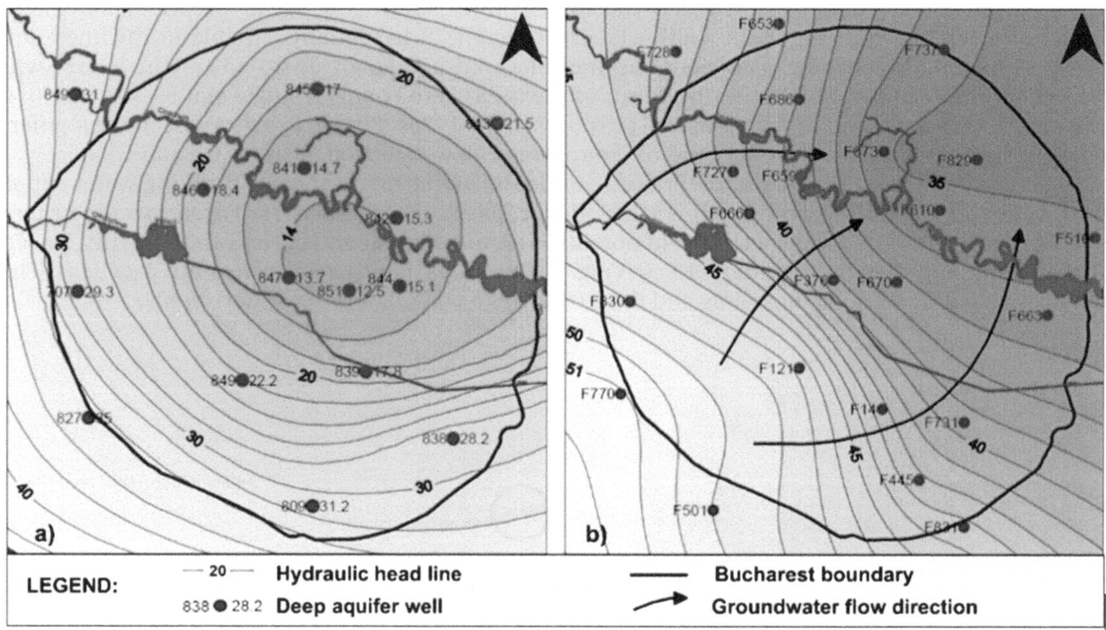

Figure C9 Change in groundwater flow direction in the Fratesti aquifer from 1981 to 2011 (after Ivan & Popa, 2012).

C5 BEBERIBE (AQUIFER)

Use of the beberibe aquifer as a strategic water-supply reserve and regulating private groundwater supply (data from Foster et al., 2010a)

The rapidly-developing Recife Metropolitan Region (RMR) has a population of over 3.0 million and a maximum water demand approaching 15 000 l/s (including high 'non-accounted for' losses exceeding 50%). It is situated in a humid tropical zone with about 2000 mm/a rainfall and is divided into two main areas: RMR-Norte and RMR Centro-Sul by a major geological lineament, leading to sharply contrasting hydrogeological conditions and groundwater potential on either side. In RMR-Norte the

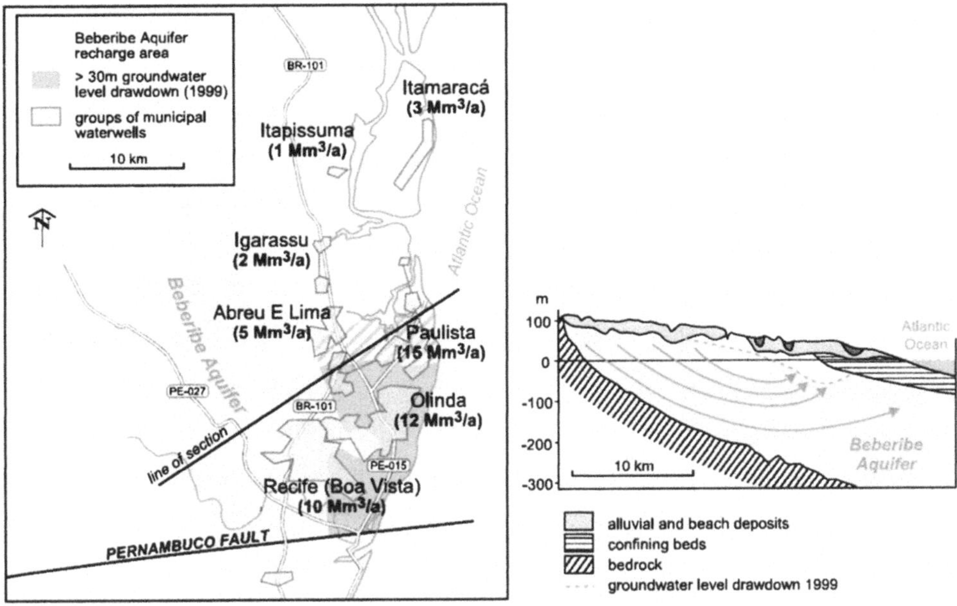

Figure C10 Distribution and structure of the Beberibe aquifer system in Recife.

Beberibe Aquifer, an Upper Cretaceous sandstone of some 200 m thickness dipping below younger strata as the coastline is approached, exhibits waterwell specific capacities of 1–2 l/s/m and is partly confined by overlying argillaceous strata up to 20 m thick (Figure C10).

COMPESA (the state water service utility) progressively developed surface water sources in four separate systems to supply the RMR area with about 8000 l/s by 2008, but this reduces in severe drought to less than 3000 l/s. However, the projected Pirapama System will provide a substantial additional supply. The operation of these systems is such that only the Botafogo System (yielding 500–1600 l/s) can supply the suburb of Olinda and northwards. The Beberibe Aquifer has been used by COMPESA to provide mains water-supply since 1975, when 22 waterwells were brought into production to provide 690 l/s. The aquifer response, with water-level drawdown locally to around −60 m MSL, gave cause for concern, but was not accompanied by rapid saline intrusion. Continued expansion of groundwater production by COMPESA occurred, and by 1985 reached a total of 1000 l/s. From 1987 development of the Botafogo System provided a potential source of 1600 l/s (allowing groundwater abstraction to be scaled back to 690 l/s), but this reduced to only 500 l/s in drought. Thus during 1990–95 groundwater production increased to 1970 l/s (from 137 operating wells), although by 2002 peak production was around 1500 l/s. In addition, it is estimated that private industrial water wells in RMR-Norte are abstracting about 700 l/s, although the depth of aquifer productive horizons is such as to have largely prevented the phenomena of residential self-supply from groundwater in this area. The Beberibe Aquifer is a good example of a high-yielding groundwater system close to a major urban area, whose geographical extension and regional flow is not sufficient for it to become a 'sole source' of urban water-supply, but whose freshwater storage reserves are large and need to be proactively protected and conjunctively managed to provide increased water-supply security at minimum possible cost (Figure C11).

In RMR-Sul, shallow groundwater is intensely exploited by some 6000–8000 private waterwells, mainly drilled in response to extreme municipal water shortages during the 1993–94 and 1999–2000 droughts, and these are still used (an estimated rate of 2000 l/s) as a low-cost supply for multi-residential

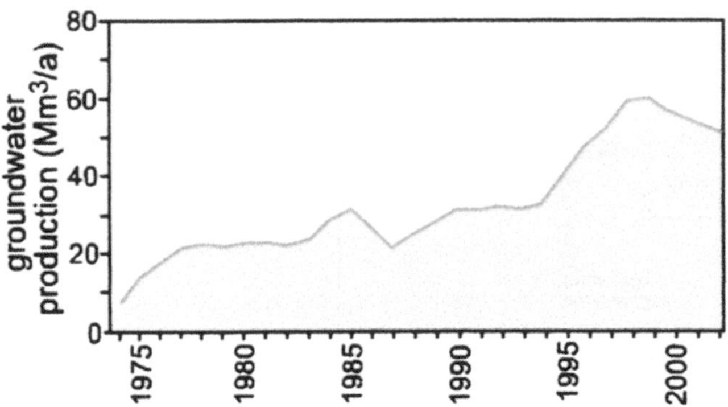

Figure C11 Evolution of groundwater production from the Beberibe aquifer in Recife.

properties and hotel facilities. Groundwater potential is much less, with high susceptibility to saline down-coning/intrusion and vulnerability to anthropogenic pollution, and uncontrolled abstraction associated with very high rates (>500 lpd) of residential water use, has caused widespread aquifer degradation. A great effort has been made to 'regularise' this private use of groundwater for domestic, commercial and industrial self-supply, by installing meters on the majority of waterwells to measure use and levy charges for discharge to the public mains sewer system in the area.

To realise an adaptive management strategy to offer greater long-term water-supply security, the following is required:
- hydrogeological investigation to identify any saline water interfaces in the aquifers, the presence of any overlying patches of polluted or saline groundwater, and the most probable aquifer recharge mechanisms and rates (including water mains leakage)
- a detailed survey and inventory of current industrial and commercial groundwater abstraction, including updating of the administrative status of use permits to arrive at a reliable estimate for non-COMPESA groundwater abstraction and its seasonality
- construction of a numerical aquifer model (using university staff working closely with COMPESA), calibrated to transient conditions with historic groundwater abstraction and drawdown data, primarily to inform dialogue about amplifying conjunctive use options through evaluation of scenarios of increased drought groundwater abstraction.

C6 LUSAKA (ZAMBIA)
Efforts to confront groundwater quality protection and meet pro-poor demand in a fast-growing city (data from Nkhuwa, 2003; Kangomba & Bäumle, 2013; and Foster *et al.*, 2020a)

Lusaka has grown rapidly from a population of 0.5 million in 1978 to 2.8 million in 2018. It has long been dependent on local groundwater for its public water-supply, as are commercial and industrial users. In 2018, the Lusaka Water and Sewerage Company (LWSC) operated 228 water wells in the public-supply network and these provided about 140 Ml/d or about 60% of their total supply, with a treatment works on the Kafue River providing a further 80 Ml/d. LWSC is still plagued by high water losses and poor revenue collection, but has taken an important 'pro-poor' initiative by drilling stand-alone boreholes to supply public water kiosks at a subsidised tariff of US$0.25/m³ (a 40–70% reduction). In addition, there are several thousand private waterwells (with total abstraction in range 80–340 Ml/d according to season), and in low-income peri-urban areas

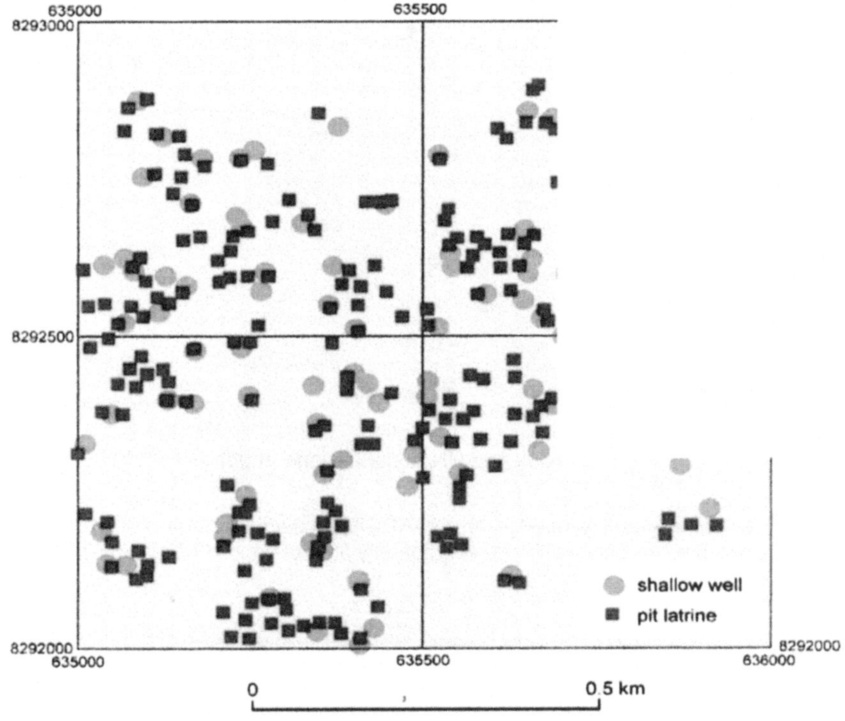

Figure C12 Close juxtaposition of domestic pit latrines and shallow private water wells in part of Lusaka.

(where 70% reside) most households still rely on shallow dug wells where the water-table is at less than 3 m depth (Figure C12).

Most wells are located within the built-up area with very little protection of their recharge capture areas, and the dolomitic-limestone formations they tap (while generally high yielding) are very vulnerable to pollution from urban wastewater, industrial effluents and agricultural cultivation. In the peri-urban areas pit latrines are the predominant form of sanitation with less than 20% of fecal matter being safely managed. The combination of unsafe sanitation, dependence on shallow wells and the karst aquifer constitute a serious hazard for groundwater quality and are the cause of frequent cholera outbreaks. Some large-scale projects to extend main sewerage and wastewater treatment are underway, but in the unplanned peri-urban areas these are difficult to implement. Additionally, innovative methods of decentralised sanitation and fecal sludge management services have been established and promoted in order to enhance the situation in the compound areas (Figure C13).

Governance, especially the coordination of water-supply, water resources protection and sanitation, has been a major challenge in Lusaka. In 2016, the Lusaka Water Security Initiative (LuWSI) was founded as a multi-stakeholder platform that aims to enhance the collaboration of key sectors such as water-supply and sanitation, urban planning, local and national authorities. Apart from awareness raising, knowledge creation and dissemination, the initiative is currently engaged in setting up and implementing groundwater protection zones for two major LWSC waterwells (Figure C14).

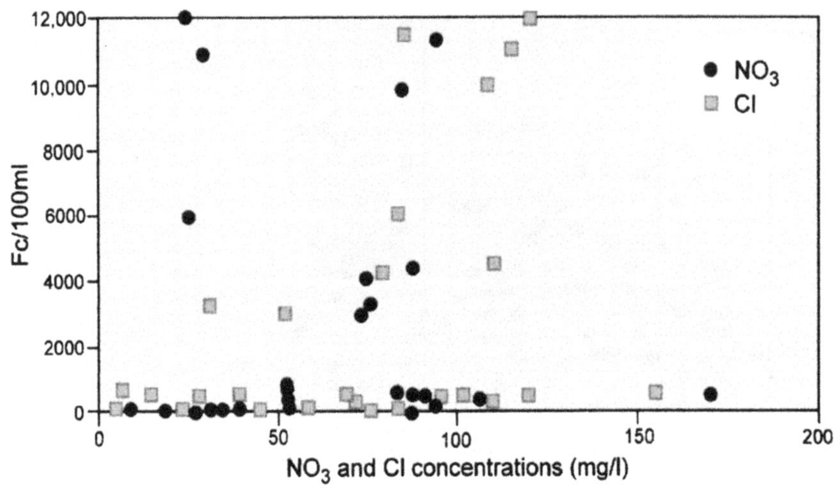

Figure C13 Lusaka waterwell fecal coliform (FC), nitrate (NO$_3$) and chloride (Cl) data (high FC counts correspond to shallow private wells but deeper utility wells are not generally affected, although can have high nitrate levels.

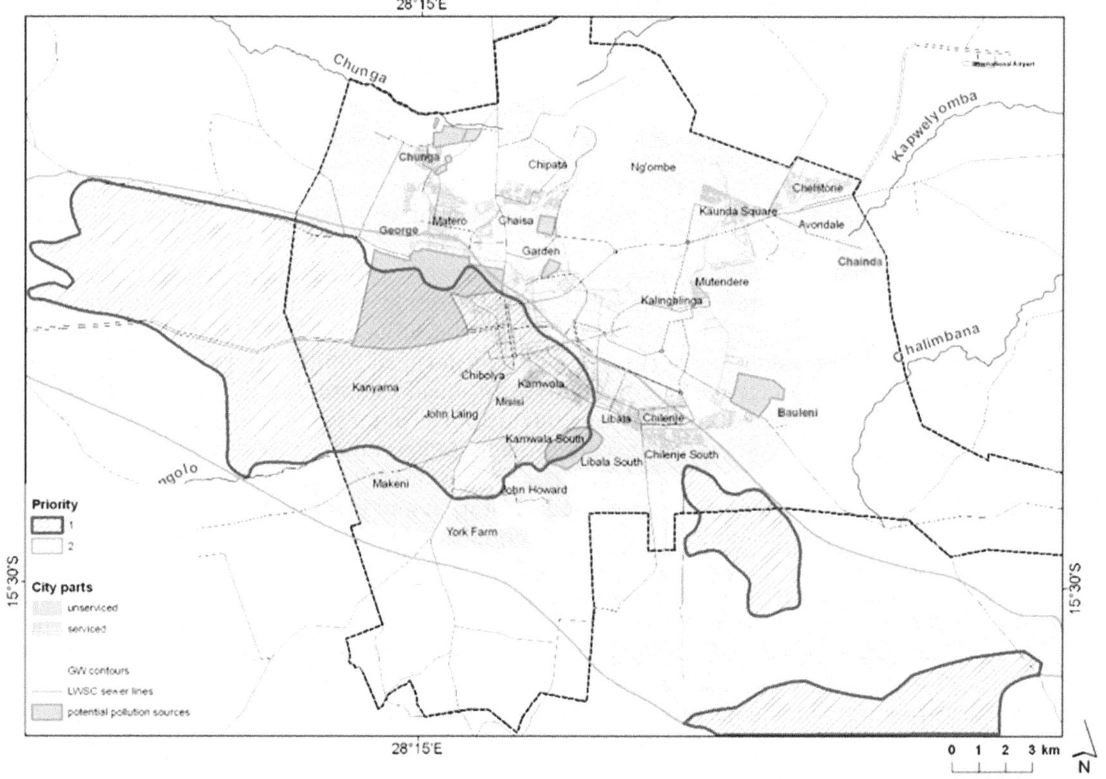

Figure C14 Priority intervention areas for groundwater protection.

References

Alam M. F. and Foster S. (2019). Policy priorities for the boom in urban private wells. IWA-The Source. October 2019, pp. 54–57.

Buapeng S. and Foster S. (2008). Controlling groundwater abstraction and related environmental degradation in Metropolitan Bangkok–Thailand. GW-MATe Case Profile Collection 20. World Bank (Washington DC). Download from www.un-igrac.org/publications (accessed 5 June 2021).

FAO-UN (2016). Groundwater Governance – Global Framework for Action to Achieve the Vision. UN Food and Agricultural Organisation (Rome) Publication, 115pp. Download from www.groundwatergovernance.org (accessed 5 June 2021).

Foster S. and Hirata R. (2011). Groundwater use for urban development: enhancing benefits and reducing risks. *Water Front*, **2011**, 21–29.

Foster S. and Vairavamoorthy K. (2013). Urban groundwater–policies and institutions for integrated management. GWP Perspectives Paper. Global Water Partnership (Stockholm).

Foster S., *et al.* (2002). Groundwater Quality Protection – A Guide for Water Utilities, Municipal Authorities and Environment Agencies. World Bank (Washington DC) Publication, 103pp. Download from www.un-igrac.org/groundwatermanagement (accessed 5 June 2021).

Foster S., *et al.* (2010a). Urban groundwater use policy–balancing the benefits and risks in developing nations. GW-MATe Strategic Overview Series 3. World Bank (Washington DC) access via www.un-igrac.org (accessed 5 June 2021).

Foster S., *et al.* (2010b). Conjunctive use of groundwater and surface water – from spontaneous coping strategy to adaptive resource management. GW-MATe Strategic Overview Series 2. World Bank (Washington DC). Download from www.un-igrac.org/groundwatermanagement (accessed 5 June 2021).

Foster S., *et al.* (2010c). Urban Groundwater Use Policy – balancing the benefits and risks in developing nations. GW-MATe Strategic Overview Series 3: 36pp. World Bank (Washington DC). Download from www.un-igrac.org/groundwatermanagement (accessed 5 June 2021)

Foster S., *et al.* (2010d). Conjunctive Use of groundwater and Surface Water – from spontaneous coping strategy to adaptive resource management. GW-MATe Strategic Overview Series 2: 26pp. World Bank (Washington DC). Download from www.un-igrac.org/groundwatermanagement (accessed 5 June 2021).

Foster S., *et al.* (2013). The aquifer pollution vulnerability concept: aid or impediment in promoting groundwater protection? *Hydrogeology Journal*, **21**, 1389–1392, https://doi.org/10.1007/s10040-013-1019-7

Foster S., *et al.* (2018). Urban groundwater use in tropical Africa- a key factor in enhancing water security? *Water Policy*, **20**, 982–994, https://doi.org/10.2166/wp.2018.056

Foster S., *et al.* (2020a). Securing the critical role of groundwater for the resilient water-supply of urban Africa. *Water Policy*, **22**, 121–132, https://doi.org/10.2166/wp.2020.177

Foster S., *et al.* (2020b). Climate change: the utility groundwater role in supply security. IWA-The Source April 2020, pp. 50–54.

Foster S., *et al.* (2020c). Urban water shortages: is groundwater the answer? CIWEM The Environment February 2020.

Gogu C. R. and Dassargues A. (2000). Current trends and future challenges in groundwater vulnerability assessment using overlay and index methods. *Environmental Geology*, **39**, 549–559, https://doi.org/10.1007/s002540050466

Grönwall J., *et al.* (2010). Groundwater, self-supply and poor urban dwellers – a review with case studies of Bangalore and Lusaka. IIED Human Settlements-Water and Sanitation Paper 26 (London).

IAH (2015). Resilient Cities and Groundwater. International Association of Hydrogeologists-Strategic Overview Series www.iah.org (accessed 5 June 2021).

IAH (2019). Climate-Change Adaptation and Groundwater. International Association of Hydrogeologists-Strategic Overview Series www.iah.org (accessed 5 June 2021).

IUCN (2016). Spring – Managing Groundwater Sustainably. International Union for Conservation of Nature (Gland) Publication 133pp. ISBN 978-2-8317-1789-0.

Ivan I. and Popa I. (2012). Up-to-date hydrodynamic and hydrochemical state of the Fratesti deep aquifer system in the Bucharest area. Scientific Bulletin. Politehnica University of Timisoara, Romania-Transactions on Hydrotechnics. TOM **57**(71), Fasc 1: pp. 13–18.

Kangomba S. and Bäumle R. (2013). Development of a groundwater information and management program for the Lusaka groundwater system: key recommendations and findings. Lusaka Department for Water Affairs and Federal Institute for Geosciences and Natural Resources (BGR) Report.

Lapworth D. J., *et al.* (2017). Urban groundwater quality in Sub-saharan Africa: current status and implications for water security and public health. *Hydrogeology Journal*, **25**, 1093–1116, https://doi.org/10.1007/s10040-016-1516-6

Nkhuwa D. C. W. (2003). Human activities and threats of chronic epidemics in a fragile geologic environment. *Physics and Chemistry of the Earth*, **28**, 1139–1145, https://doi.org/10.1016/j.pce.2003.08.035

Thomsen R., *et al.* (2004). Hydrological mapping as a basis for establishing site-specific groundwater protection zones in Denmark. *Hydrogeology Journal*, **12**, 550–562, https://doi.org/10.1007/s10040-004-0345-1

Tuinhof A. *et al.* (2006). Groundwater Monitoring Requirements – for managing aquifer response and quality threats. GW-MATe Briefing Note Series 9: 10. World Bank (Washington DC). Download from www.un-igrac.org/groundwatermanagement (accessed 5 June 2021).

IWA Publishing's authorised EU representative for General Product
Safety Regulations is Diane D'Arras, 15 rue Duret, 75116 Paris,
France, e-mail: safety@iwap.co.uk.

Printed and bound by CPI Group (UK) Ltd, Croydon, CR0 4YY
18/05/2026
02110092-0003